My Book

This book belongs to

Name: ______________________________

MATH KNOTS
MK

Cover Design by :
Gowri Vemuri

First Edition :
January , 2020

Second Edition : January , 2023

Author :
Gowri Vemuri

Edited by :
Ritvik Pothapragada
Siddhartha Rangavajjula

MATH KNOTS
Joy of Learning...

Dedication

This book is dedicated to:

My Mom, who is my best critic, guide and supporter.

To what I am today, and what I am going to become tomorrow,

is all because of your blessings, unconditional affection and support.

This book is dedicated to the

strongest women of my life ,

my dearest mom

and

to all those moms in this universe.

G.V.

MK
MATH KNOTS
Joy of Learning...

American Computer Science League (ACSL) is an international computer science competition originally founded in 1978. This organization is also an institutional member of the Computer Science Teachers Association. ACSL is on the approved activities list of the National Association of Secondary School Principals (NASSP).

ACSL consists of five divisions to appeal to the varying computing abilities and interests of students. All students at a school can take the tests but can only participate in one division. A team score is the sum of the best 3 or 5 scores in each test. Those scores can come from different students with in the team, in each contest. Prizes are awarded to top scoring students and teams based on cumulative scores after the 4th test.

The **Senior / Intermediate Division** is geared to those high school students with programming experience. Each contest consists of a 30-minute, 5- questions short answer test and a take home programming problem to be solved in 72-hours. Team scores can be based on the sum of the top 3 or top 5 scores each contest.

The **Junior Division** is geared to junior high and middle school students with no previous experience programming computers. No student beyond grade 9 may compete in the Junior Division. Each contest consists of a 30-minute 5-question short answer test and a take home program to be solved in 72-hours. Team scores are based on the sum of the best 5 scores each test.

The **Classroom Division** is open to students from all grades. It consists of a selection of the non-programming problems. As its name implies, this division is particularly well-suited for use in the classroom. Each contest consists of a 50-minute, 10-question short answer test. Team scores are based on the sum of the best 5 scores each test.
The **Elementary Division** is open to students from grades 3 - 6. It consists of non-programming problems. Four categories, one each contest, will be tested. The contest consists of a 30-minute, 5-questions test.

ELEMENTARY DIVISION	CLASS ROOM / JUNIOR DIVISION	INTERMEDIATE / SENIOR DIVISI
Elementary Computer Number Systems	Computer Number Systems	Computer Number Systems
Elementary Prefix/Infix/Postfix Notation	Recursive Functions	Recursive Functions
Elementary Boolean Algebra	What Does This Program Do? Branching	What Does This Program Do?
Elementary Graph Theory	Prefix/Infix/Postfix Notation	Prefix/Infix/Postfix Notation
	Bit-String Flicking	Bit-String Flicking
	What Does This Program Do? Loops	LISP
	Boolean Algebra	Boolean Algebra
	Data Structures	Data Structures
	What Does This Program Do? Arrays	FSA/Regular Expressions
	Graph Theory	Graph Theory
	Digital Electronics	Digital Electronics
	What Does This Program Do? Strings	Assembly Language

This book is written by computer science teachers and industry experts.
Our book comprises of various practice questions in line with ACSL topics.

NOTE: ACSL is neither affiliated nor endorsed the content of this book.

This book is intended to give a hands on practice on various topics covering contest 1 and contest 2.
The questions might not replicate 100% of the actual test but are intended to reinforce the basic concepts.

Index

MK
MATH
KNOTS
Joy of Learning...

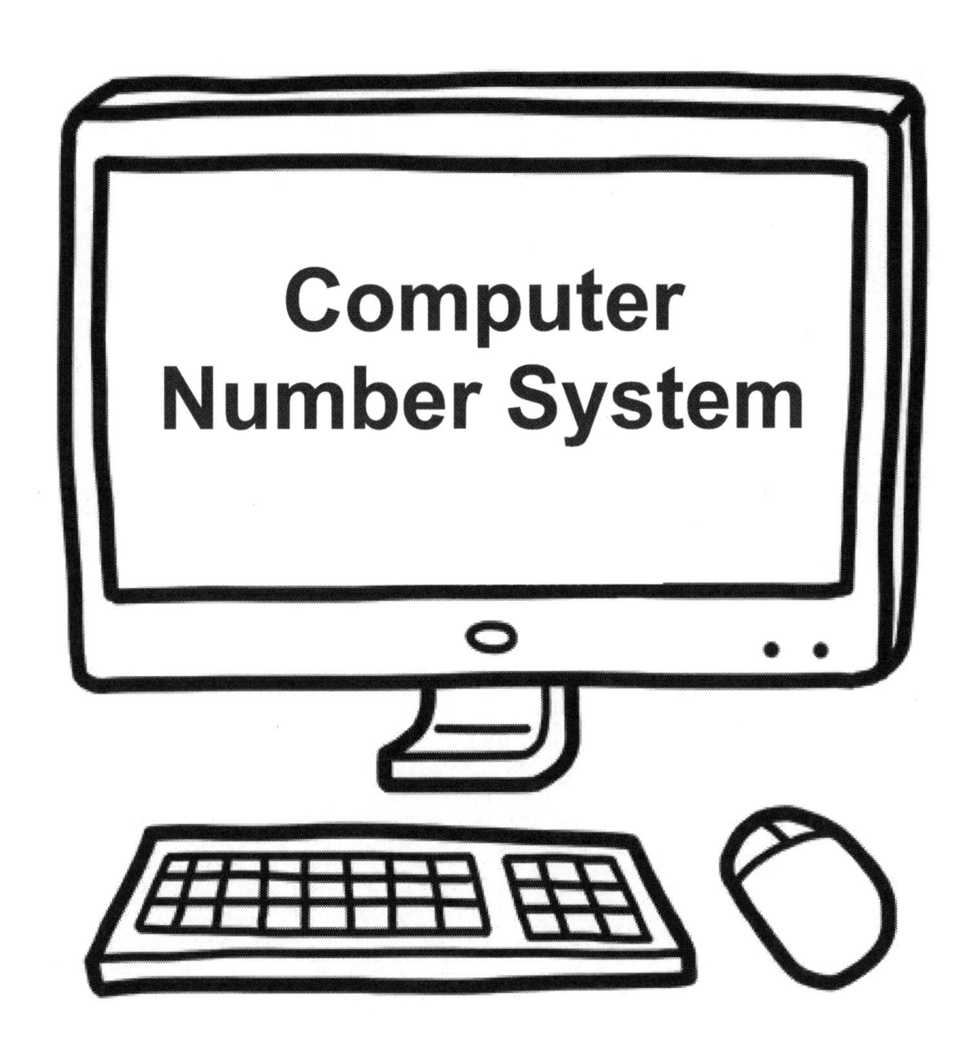
Computer
Number System

MK
MATH
KNOTS

NUMBER SYSTEMS

- Decimal
- Binary
- Hexadecimal
- Octal

Decimal System:

Any number which is positionally defined in the number system is called as a decimal. All decimal numbers are with base 10.

Example :245

$= 2 \times 10^2 + 4 \times 10^1 + 5 \times 10^0 = 200 + 40 + 5$

Binary System:

As the name implies binary numbers are only two : 1 and 0 . All binary numbers are represented with base 2

Example: Convert the following to decimal system representation 0110100

64	32	16	8	4	2	1
0	1	1	0	1	0	0

$= 0X2^6 + 1X2^5 + 1X2^4 + 0X2^3 + 1X2^2 + 0X2^1 + 0X2^0$

= 0X64 + 1X32 + 1X16 + 0X8 + 1X4 + 0X2 + 0X2

= 0 + 32 + 16 + 0 + 4 + 0 + 0 = 52

Octa decimal System:

All octa decimal numbers are represented with base 8

Hexa decimal System: All octa decimal numbers are represented with base 16

Comparison chart of various Number Systems

Decimal	Binary	Octal	Hexadecimal
0	0000	0	0
1	0001	1	1
2	0010	2	2
3	0011	3	3
4	0100	4	4
5	0101	5	5
6	0110	6	6
7	0111	7	7
8	1000	10	8
9	1001	11	9
10	1010	12	A (10)
11	1011	13	B (11)
12	1100	14	C (12)
13	1101	15	D (13)
14	1110	16	E (14)
15	1111	17	F (15)

Grouping Binary Digits for Octal and Hexadecimal Numbers

EXAMPLE:

BINARY TO OCTAL

111000111_2 = 111000111 = 707_8

BINARY TO HEXA DEECIMAL

111000111_2 = 111000111 = $1C7_{16}$

COLOR REPRESENTATION :

All computer graphics are created using three main colors Red, Green and Blue. A combination of these produce various shades of the colors that we see on computer screens/graphics. Every color is represented by 6 digits with a hash sign preceding them. First two digits represent color Red, next two digits represent Green and last two digits represent Blue

White is represented by : #FFFFFF
in which first two FF is represented by Red ,
next two FF is represented by Green ,
last two FF are represented by B lue

Red Color is always represented by #FF0000
BlueColor is always represented by #00FF00
GreenColor is always represented by #0000FF

Example

FFE4B5

Red Color is represented by FF = 255

Blue Color is always represented by E4 = 228

Green Color is always represented by B5 = 181

Find the decimal values of red ,blue and green colors in $BDB76B_{16}$

BD_{16}= 11*16 + 13*1 = 189

$B7_{16}$ = 11*16 + 7*1 =183

$6B_{16}$ = 6*16 + 11*1 =107

1. Convert 120_8 to base 10.

2. Convert 155_8 to base 10.

3. Convert 236_8 to base 10.

4. Convert 334_8 to base 10.

5. Convert 606_8 to base 10.

6. Convert 328_8 to base 10.

7. Convert 1121_8 to base 10.

8. Convert 2156_8 to base 10.

9. Convert 2177_8 to base 10.

10. Convert 32009_8 to base 10.

11. Convert 1079_8 to base 10.

12. Convert $2AB_{16}$ to base 10.

13. Convert $3AD_{16}$ to base 10.

14. Convert $3CE_{16}$ to base 10.

15. Convert $2DE_{16}$ to base 10.

16. Convert $1AF_{16}$ to base 10.

17. Convert $3DA_{16}$ to base 10.

18. Convert AAF_{16} to base 10.

19. Convert $A5E_{16}$ to base 10.

20. Convert $A3F_{16}$ to base 10. .

21. Convert $38C_{16}$ to base 10.

22.Convert the following base 8 number to base 16:

24675_8

23. Convert the following base 8 number to base 16:

44453_8

24. Convert the following base 8 number to base 16:

31722_8

25. Convert the following base 8 number to base 16:

50116_8

26.Convert the following base 8 number to base 16:

25732_8

27. Convert the following base 8 number to base 16:

204452_8

28. Convert the following base 8 number to base 16:
150147_8 .

29. Convert the following base 8 number to base 16:
222206_8 .

30. Convert the following base 8 number to base 16:
303033_8

31. Convert the following base 8 number to base 16:

127244_8

32. Convert A52DE_{16} to Octal.

33. Convert DAC 5A_{16} to Octal.

34. Convert $F785D_{16}$ to Octal

35. Convert $C980F_{16}$ to Octal

36. Convert $B13B7_{16}$ to Octal

37. Convert $ED0B8_{16}$ to Octal

38. Convert $C45A2E_{16}$ to Octal

39. Convert $AA5DE_{16}$ to Octal

40. Convert $CC13EE_{16}$ to Octal

41. Convert $FEFE5F_{16}$ to Octal

42. How many more 1's are there in the binary representation of $5B4_{16}$ than in the binary representation of $C41_{16}$?

43. How many more 1's are there in the binary representation of $7BE1_{16}$ than in the binary representation of $10D4_{16}$?

44. How many more 1's are there in the binary representation of $F4D8_{16}$ than in the binary representation of $18DE_{16}$?

45. How many more 1's are there in the binary representation of $CCD5_{16}$ than in the binary representation of $14AF_{16}$?

46. How many more 1's are there in the binary representation of $91FE_{16}$ than in the binary representation of $88C5_{16}$?

47. Express the sum in hexadecimal:

$54D1E_{16} + 1A2D_{16}$

48. Express the sum in hexadecimal:

$27E_{16} + 7CA_{16}$

49. Express the sum in hexadecimal:

$7AC_{16} + 8BD_{16}$

50. Express the sum in hexadecimal:

$5EA_{16} + 6FC_{16}$

51. Express the sum in hexadecimal:

$68F_{16} + 8FB_{16}$

52. Simplify the below in hexadecimal:

$B7F_{16} - 5DB_{16}$

53. Simplify the below in hexadecimal:

$DE2_{16} - 3EA_{16}$

54. Simplify the below in hexadecimal:

$FF8_{16} - AAA_{16}$

55. Simplify the below in hexadecimal:

CFD_{16} - $AB3_{16}$

56. Simplify the below in hexadecimal:

$DA9_{16}$ - $1DC_{16}$

57. Express the sum in octa decimal:

$449_8 + 556_8$

58. Express the sum in octa decimal:

$5327_8 + 2117_8$

59. Express the sum in octa decimal:

$4122_8 + 3588_8$

60. Simplify in octa decimal:

$6565_8 - 3985_8$

61. Simplify in octa decimal:

$5904_8 - 1288_8$

62. Which is larger from the below, Order them from least to greatest

(A) 120_8 (B) 334_8 (C) 328_8

63. Which is larger from the below, Order them from least to greatest

(A) $3AD_{16}$ (B) 236_8 (C) $2DE_{16}$

64. Which is larger from the below ,Order them from least to greatest

(A) 606_8 (B) $3DA_{16}$ (C) $1CE_{16}$

65. Which is larger from the below, Order them from least to greatest

(A) $2AB_{16}$ (B) 2156_8 (C) 1121_8

66. Which is larger from the below, Order them from least to greatest

(A) 1079_8 (B) AAF_{16} (C) $38C_{16}$

67) What is the decimal value of the green component in the RGB code?

$6B8E23_{16}$

68) Is the decimal value of the blue and green component in the RGB code given below are equal is so what are the values?

$48D1CC_{16}$

69) What are the decimal values of the Red, Green and Blue components in the RGB code?

$1E90FF_{16}$

70) Is the decimal value of the blue and green component in the RGB code given below are equal is so what are the values, Which one is greater and by what value?

$F5DEB3_{16}$

71) Is the decimal value of the blue and green component in the RGB code given below are equal is so what are the values, Which one is greater and by what value?

$A0522D_{16}$

Prefix - Postfix

MK
MATH KNOTS
Joy of Learning...

Prefix-Postfix Notations

Infix Notation:

P + Q

As a general way of writing expressions, Operators are written in-between their operands.

An expression such as A *(B + C) /D

is simplified as "First add B and C together, then multiply the result by A, then divide by D."

The Infix Notations use the general order of evaluation, using PEDMAS rules

Prefix Notation:

Infix, Postfix and Prefix notations are three different but equivalent ways of writing expressions. It is easiest to demonstrate the differences by looking at examples of operators that take two operands.

Infix notation: A + B.　　　　Prefix notation: + X Y

Operators are written before their operands.

An expression such as A *(B + C) /D = / (A *(B + C)) D

= / * A(B+C) D /= / * A + B C D

- The order of evaluation of operators is always left-to-right Brackets cannot be used to change this order.
- Because the "/" is to the left of the "*" in the example above, the addition must be performed before the multiplication.
- Operators act on values immediately to the left of them. For example, the "+" above uses the "B" and "C".
- Brackets can be added to make this more explicit
  (/ (* A (+ B C)) D)

SAMPLE #1:

Convert the following infix to prefix notation:

$$(A - D) + \frac{B}{C}$$

$= ((A - D)) + (\frac{B}{C})$

$= ((-AD)) + (\frac{B}{C})$

$= (-AD) + (/BC)$

$= + - AD/BC$

Postfix Notation:

Infix notation: A + B

Postfix notation (also known as "Reverse Polish notation"): A B +

Operators are written after their operands.

An expression such as A *(B + C) /D = A *(B + C) D /

= A(B+C) *D / = A B C + * D /

- The order of evaluation of operators is always left-to-right
- Brackets cannot be used to change this order.
- Because the "+" is to the left of the "*" in the example above, the addition must be performed before the multiplication.
- Operators act on values immediately to the left of them. For example, the "+" above uses the "B" and "C".
- Brackets can be added to make this more explicit

  ((A (B C +) *) D /)

Thus, the "*" uses the two values immediately preceding: "A", and the result of the addition. Similarly, the "/" uses the result of the multiplication and the "D".

In all notations, the operands occur in the same order, and just the operators have to be moved to keep the meaning correct.

SAMPLE #2:

Convert the following infix into postfix notations:

$$\frac{\boldsymbol{A+B+C\ +A^2}}{\mathbf{4}}$$

$$=\frac{((A+B+C)+A^2)}{4}$$

$$=\frac{((AB+C)+A^2)}{4}$$

$$=\frac{(ABC+A2\uparrow+)}{4}$$

$$=ABC+A2\uparrow+4/$$

MATH KNOTS
Joy of Learning...

1.Evaluate the following postfix expression. Note that all numbers are not single digit.

25 25 / 72 * 8 − +

2. Evaluate the following postfix expression. Note that all numbers are single digit.

6 5 * 3 3 + / + 3 3 * 2 -

3. Evaluate the following postfix expression. Note that all numbers are not single digit.

   15 3 / 5 9 * 5 − +

4. Evaluate the following postfix expression. Note that all numbers are not single digit.

   88 8 /8 8 * 8 − +

5. Evaluate the following postfix expression. Note that all numbers are not single digit.

   65 5 / 7 2 * 7 − +

6. Evaluate the following postfix expression. Note that some numbers are not single digit.

   9 4 * 9 9 + / 10 4 * + 5 -

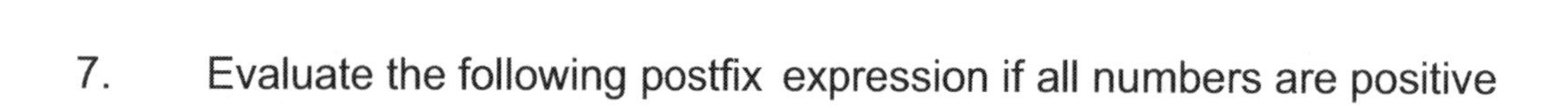

7. Evaluate the following postfix expression if all numbers are positive

   6 3 4 + 1 ^ * 12 2 / 4 * -

8. Evaluate the following postfix expression if all numbers are single digits.

   7 1 1 + 2 ^ * 14 7 / 7 * -

9. Evaluate the following postfix expression if all numbers are single digits.

2 2 2 + 2 ^ * 2 2 / 2 * -

10. Evaluate the following postfix expression if all numbers are positive

8 48 + 2 ^ * 32 8 / 4 * -

11. Evaluate the following postfix expression if all numbers are not single digits.

10 6 2 - 3 ^ * 50 5 / 4 * +

12. Evaluate the following prefix expression. Note that all numbers are single digits in the expression.

+ / + 5 5 5 ^ 5 1

13. Evaluate the following prefix expression if all numbers are not single digits.

– * + 22 11 3 * 7 / 15 5

14. Evaluate the following prefix expression. Note that all numbers are single digits in the expression.

– * + 6 4 6 ^ 3 1

15. Evaluate the following prefix expression if all numbers are single digits.

    * * + 1 1 1 * 1 / 4 2

16. Evaluate the following prefix expression. Note that all numbers are single digits in the expression.

    - * + 5 6 5 ^ 5 1

17. Evaluate the following prefix expression if all numbers are single digits.

    + * - 15 4 5 + 4 / 8 8

18. Evaluate the following prefix expression. Note that all numbers are single digits in the expression.

    * ^ / 2 2 2 + 2 2

19. Evaluate the following prefix expression if all numbers are single digits.

    + ^ -7 0 2 * 4 / 6 2

20. Evaluate the following prefix expression. Note that all numbers are single digits in the expression.

    ^ * / + 4 4 4 ^ 4 2 0

21. Evaluate the following prefix expression if all numbers are single digits.

    + * + 8 2 8 * 4 / 8 2

22. Evaluate the following prefix expression if all numbers are not single digits.

    + * ^ 99 0 5 ^ 6 / 4 2

23.Convert the infix expression to a postfix expression. Perform all operations in the same order as the given expression.

5 * 4 / 1 - 3 ^ 3 + 7 * 2 - 4 ^ 2

24. Convert the infix expression to a postfix expression. Perform all operations in the same order as the given expression.

(9 - 5) / 1 * 5 / (9 - 3)

25. Convert the infix expression to a postfix expression. Perform all operations in the same order as the given expression.

    7- 8 * 5 + 4 * 4 - 4

26. Convert the infix expression to a postfix expression. Perform all operations in the same order as the given expression.

    $(6 + 4)^3 * (9 - 5)^2 / 4$

27. Convert the infix expression to a prefix expression. Perform all operations in the same order as the given expression.

    7 + 5 * 2 − 5 * 3 + 9

28. Convert the infix expression to prefix. Perform all operations in the same order as the expression.

    9 / (6 − 3) + (4 + 5)^3 − 5 * 7

29. Convert the infix expression to a prefix expression. Perform all operations in the same order as the given expression.

$$5^{(2 + 3)} * (9 - 1)^3$$

30. Convert the infix expression to a prefix expression. Perform all operations in the same order as the given expression.

$$3^{(3 + 3)} / (7 - 4)^3$$

31. Translate the following postfix expression to prefix if all numbers are single digits.

9 8 + 7 * 9 7 8 + * −

32. Translate the following postfix expression to prefix if all numbers are single digits.

4 3 − 5 4 + 3 * 5 3 4 + * −

33. Translate the following postfix expression to prefix if all numbers are single digits.

6 9 + 3 / 6 3 1 + * −

34. Translate the following postfix expression to prefix if all numbers are single digits.

7 7 + 7 / 1 7 7 + * −

35. Translate the following prefix expression to postfix. Note that all numbers are single digits in the expression.

    * ^ + 7 4 9 – 8 1

36. Translate the following prefix expression to postfix. Note that all numbers are single digits in the expression.

    / ^ + 1 2 5 – 5 2

37. Translate the following prefix expression to postfix. Note that all numbers are single digits.

    * / – 6 2 4 * + 3 1 + 5 3

38. Translate the following prefix expression to postfix. Note that all numbers are single digits.

    * / + 5 5 1 * – 5 1 * 1 3

39. Translate the following prefix expression to postfix. Note that all numbers are single digits.

+ / − 6 1 6 * + 1 6 1

40. Translate the following prefix expression to postfix. Note that All numbers are single digits.

+ / − 4 3 4 * + 3 4 3

41. Translate the following prefix expression to postfix. Note that all numbers are single digits.

    + / + 4 7 6 * * 7 6 4

42. Translate the following prefix expression to postfix. Note that all numbers are single digits.

    + / − 7 3 7 * * 7 3 7

43.Translate the following prefix expression to postfix. Note that all numbers are single digits.

– / – 2 1 7 * * 2 2 1

44. Translate the follow postfix expression to prefix. Note that all numbers are single digits.

3 3 * 2 2 + – 3 2 * 2 – +

45. Translate the follow postfix expression to prefix. Note that all numbers are single digits.

    9 1 * 5 2 + - 6 5 * 6 - -

MATH KNOTS

Boolean Algebra

MATH KNOTS

Boolean Algebra

Boolean Algebra is the mathematics we use to analyze digital gates and circuits. We can use these "Laws of Boolean" to both reduce and simplify a complex Boolean expression in an attempt to reduce the number of logic gates required.
Boolean Algebra is a system of mathematics based on logic that has its own set of rules or laws which are used to define and reduce Boolean expressions.
The variables used in **Boolean Algebra** only have one of two possible values, 0" and "1" but an expression can have an infinite number of variables.
Each variable can have a value of 1 or 0 only.

Laws of Boolean Algebra :

1. Annulment Law – A term AND´ed with a "0" equals 0 or OR´ed with a "1" will equal 1

→ A . 0 = 0 A variable AND'ed with 0 is always equal to 0
→ A + 1 = 1 A variable OR'ed with 1 is always equal to 1

2. Identity Law – A term OR´ed with a "0" or AND´ed with a "1" will always equal that term

→ $A + A = A$
→ $AA = A$
→ A + 0 = A A variable OR'ed with 0 is always equal to the variable
→ A . 1 = A A variable AND'ed with 1 is always equal to the variable

3. Idempotent Law – An input that is AND´ed or OR´ed with itself is equal to that input

→ A + A = A A variable OR'ed with itself is always equal to the variable
→ A . A = A A variable AND'ed with itself is always equal to the variable

4. <u>Complement Law</u> – A term AND´ed with its complement equals “0” and a term OR´ed with its complement equals “1”

→ $\overline{A} . A = 0$ A variable AND'ed with its complement is always equal to 0
→ $\overline{A} + A = 1$ A variable OR'ed with its complement is always equal to 1

5. <u>Commutative Law</u> – The order of application of two separate terms is not important

→ $A . B = B . A$ The order in which two variables are AND'ed makes no difference

→ $A + B = B + A$ The order in which two variables are OR'ed makes no difference

6. <u>Double Negation Law</u> – A term that is inverted twice is equal to the original term

→ $A = A$ A double complement of a variable is always equal to the variable

7. <u>de Morgan´s Theorem</u> – There are two “de Morgan´s” rules or theorems,

→ Two separate terms NOR´ed together is the same as the two terms inverted (Complement) and AND´ed for example: $A+B = A . B$

8. **De Morgan's Theorem**

$\overline{(A + B)} = \bar{A}\,\bar{B}$

→ Two separate terms NAND´ed together is the same as the two terms inverted (Complement) and OR´ed for example: A.B = A + B

$\bar{A}\,\bar{B} = \overline{(A + B)}$

Other algebraic Laws of Boolean not detailed above include:

9. <u>Distributive Law</u> – This law permits the multiplying or factoring out of an expression.

→ A(B + C) = A.B + A.C (OR Distributive Law)
→ A + (B.C) = (A + B).(A + C) (AND Distributive Law)

10. <u>Absorptive Law</u> – This law enables a reduction in a complicated expression to a simpler one by absorbing like terms.

→ A + (A.B) = A (OR Absorption Law)
→ A(A + B) = A (AND Absorption Law)

11. <u>Associative Law</u> – This law allows the removal of brackets from an expression and regrouping of the variables.

→ A + (B + C) = (A + B) + C = A + B + C (OR Associate Law)
→ A(B.C) = (A.B)C = A . B . C (AND Associate Law)

12. $AB + A\overline{B} = A$
    $(A + B)(A + \overline{B}) = A$

13. $A + AB = A$
    $A(A + B) = A$

14. $A + \overline{A}B = (A + B)$

    $A(\overline{A} + B) = AB$

15. $A \oplus B = A\overline{B} + \overline{A}B$
    The *xor* of two values is true whenever the values are different. It uses the $\oplus$ operator, and can be built from the basic operators:

    $A \oplus B = A\overline{B} + \overline{A}B$
    The values of *xor* for all possible inputs are shown in the truth table below

A	B	$A \oplus B$
0	0	0
0	1	1
1	0	1
1	1	0

16. The *xnor* of two values is true whenever the values are the same. It is the *not* of the *xor* function. It uses the $\odot$ operator: $A \odot B = \overline{A \oplus B}$
The *xnor* can be built from basic operators:
$A \odot B = \overline{A \oplus B} = AB + \bar{A}\bar{B}$
The values of *xnor* for all possible inputs is shown in the truth table below:

A	B	$A \odot B$
0	0	1
0	1	0
1	0	0
1	1	1

Example :
Simplify the following expression: (A + B)(A + C)

(A + B)(A + C) = A.A + A.C + A.B + B.C (Distributive law)

= A + A.C + A.B + B.C (Idempotent AND Law)

= A(1 + C) + A.B + B.C (Distributive law)

= A.1 + A.B + B.C (Identity OR law)

= A(1 + B) + B.C (Distributive law)

= A + B.C (Identity AND law)

(A + B)(A + C) = A + B.C

MATH KNOTS

Evaluate the following expression as either TRUE or FALSE:

1. (((4 + 5) <= 11) AND ((9 - 5)>= 3 * 2)) OR NOT (7 * 7 < 5 * 5)

2. ((22/2) >= 22) AND (2 * 5)>= 2 * 2) OR NOT (3 + 7 >= 2 * 5)

3. ((10 - 1) <= (10 + 1) AND ((3 * 9)>= 3 * 6)) OR NOT (10 * 7 > 6 * 9)

4. ((7 + 3) <= 4 + 5) AND ((6 + 1)>= 4 * 1) OR NOT (99/11 > = 3 * 3)

5. ((1/5 <= 1/4) AND ((2/3) >= 2 * (1/3))) OR NOT (13 + 5 = 5 + 13)

6. NOT (6^2 < 4 * 6) OR NOT (10/ 5 ≥ 3 AND 12 * 4 > 10 * 5)

7. NOT (3^3 < 5^2) OR (10/ 5 ≥ 3 AND 12 * 4 > 10 * 5)

8. NOT (19-12 < = 3) OR (64 / 8 ≥ 4 * 2 AND 6 * 3 > 3 * 1)

9. NOT (10^2 < 2 ^4) OR (72 / 6 ≥ 12 AND 40 * 2 > 9 * 8)

11. How many ordered pairs make the following statement FALSE?

NOT A OR (A AND (A AND B))

12. (A OR B) AND (A OR NOT B)

13. A OR NOT (A AND B)

14. A AND NOT (A OR B)

15. **NOT A AND (A AND(A OR B))**

16. NOT A AND (A OR (A OR B))

17. ((B AND NOT NOT B) OR (A OR NOT NOT A)) OR (A AND NOT B)

18. (A OR B) OR (B AND A) AND (A AND B)

19. (A AND B) OR (B AND A) AND (A OR B)

20. (A AND B) OR (B AND NOT B) AND (A AND NOT NOT A)

21. (A AND B) AND (B AND NOT B) AND (A AND NOT NOT A)

22. (A AND B) OR (B AND NOT B) OR (B AND NOT NOT B)

23. (A OR B) AND (B OR NOT B) OR (B OR NOT NOT B)

24. NOT ((A AND A) AND (B OR B)) OR NOT (A AND B)

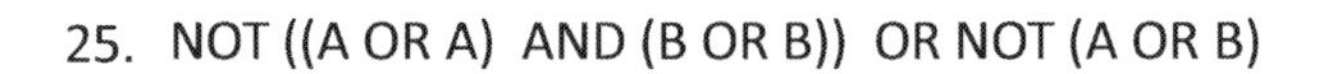

25. NOT ((A OR A) AND (B OR B)) OR NOT (A OR B)

26. (NOT A OR A) AND (B OR NOT B) OR NOT (A AND B)

27. NOT   (NOT A OR A) AND (B OR NOT B) OR NOT A

28. (B OR NOT A) AND (A OR NOT A) AND (B OR NOT B)

29. (A OR NOT NOT B) OR (A OR NOT NOT A) AND (A AND B)

30. (A OR NOT NOT B) OR (A OR NOT NOT A) OR (NOT A AND B)

31. (A + ~ B)(A + B) + (~A) + ~ A

32. ~ A~B (((A * B) * (A * B)) + (A * B))

33. A (~AB + B) + ~A~B

34. ~[~B (~AB + B) + ~A (A + ~A) + A]]

35. ~A[A(~A + B) + A .~B]

36. B (A + ~B) + ~A B + ~ B

37 ~[A (B + ~B) (A + ~A)] + ~B

38. BA + A(A + ~B) + ~AB + AB

39. B [~B~A + ~A(A + ~B) + A~B] = B [~B~A + ~AA + ~A~B + A~B]

40. Is the below statement TRUE or FALSE?

A [~A ~B + B (~A + ~B)] = A [~A ~B + B ~A + B ~B]

41. B (A + ~B) (~A + ~B)

42. (~A + B) (A + ~B)

43. ~ (~A + ~B) (A + B)

44. B[~B (B + ~B) (A + ~A)] + (~~A)(~A)

45. ~ (((A * B) * (A * B))+ (A * B)) + ~AB + A

MK
MATH KNOTS
Joy of Learning...

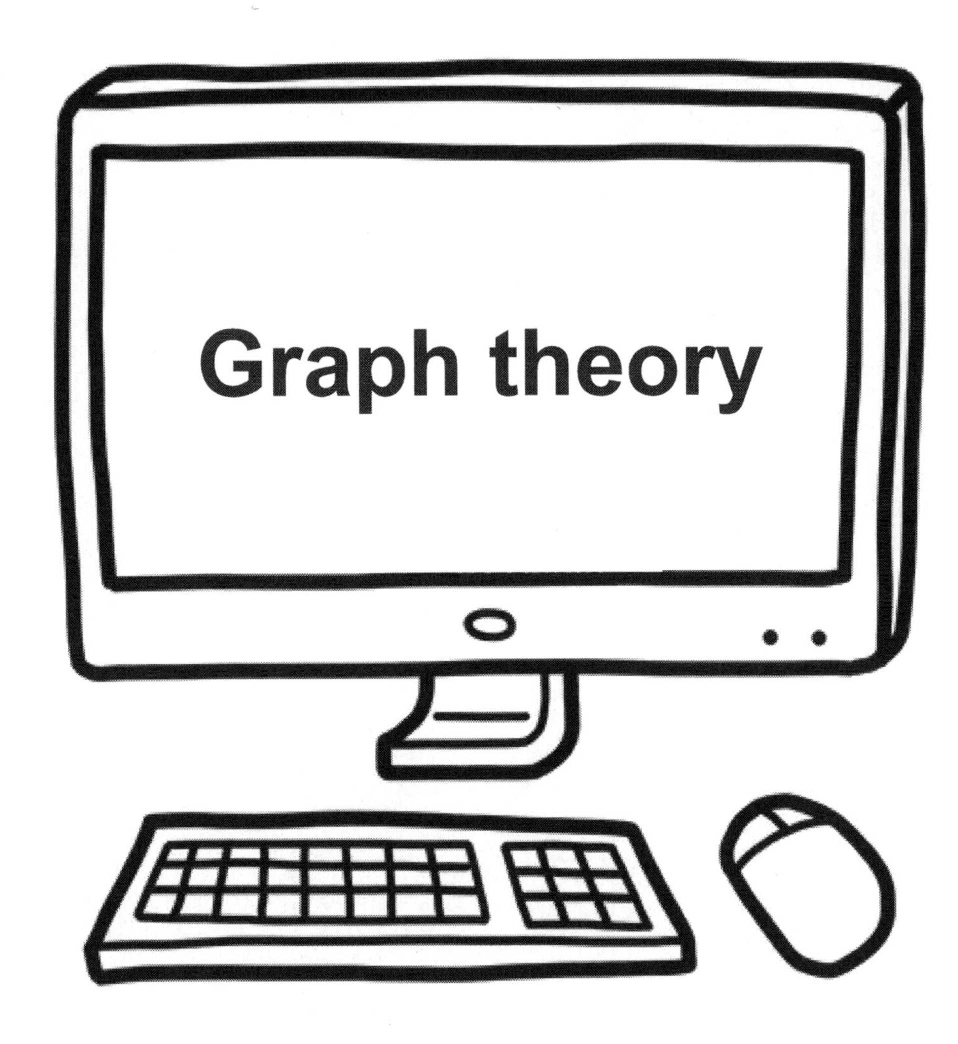
Graph theory

MATH KNOTS
MK
Joy of Learning...

GRAPH THEORY

Graph theory is used in many situations in a day to day life

Example:

#1 Minimum cost of a underground cable wire to various locations from a central location. #2 Fastest route from national capital to each state of the capital.

#3 Asports league scheduled into minimum number of weeks

A graph has vertices's and edges. A loop is an edge whose endpoints are equal.
A simple graph is a graph having no loops or multiple edges

DATA Structures

Traversable Network:

Traversability of a graph refers to weather or not you can use every edge once and only once without lifting your pencil. The most famous traversability example is the seven bridges of Konigsberg

A network is traversable if it can be drawn by tracing each edge exactly once, without lifting your pencil. If there are no odd nodes or if there are two odd nodes, that means the network is traversable

MATH KNOTS

1) Using the first letters of the names , Draw a graph relating to their meetups.
Amy can go to Bob's house , Amy can go to Cathy's house ,
Amy can go to Ella's house , Amy can go to Geeth's house ,
Geeth can go to Fiona's house , Fiona can go to Ella's house ,
Ella can go to Dan's house , Bob can go to Cathy's house ,
and Cathy can go to Dan's house.

2) Using the first letters of the names , Draw a graph relating to their meetups.
Amy can go to Bob's house , Amy can go to Cathy's house ,
Amy can go to Ella's house , Amy can go to Geeth's house ,
Geeth can go to Fiona's house , Fiona can go to Ella's house ,
Ella can go to Dan's house , Bob can go to Cathy's house ,
Cathy can go to Dan's house , Geeth can go to Cathy's house ,
and Fiona can go to Cathy's house

3) Using the first letters of the names , Draw a graph relating to their meetups.
Amy can go to Bob's house , Amy can go to Cathy's house ,
Amy can go to Ella's house , Amy can go to Geeth's house ,
Geeth can go to Fiona's house , Fiona can go to Ella's house ,
Ella can go to Dan's house , Bob can go to Cathy's house ,
Cathy can go to Dan's house , Geeth can go to Cathy's house ,
Fiona can go to Cathy's house , and Dan can go to Geeth's house

4) Zoom Zoom carnival has six crazy rides. There are only few paths to reach to one ride to another ride. Jade can go from ride A to E , ride A to B , ride A to F , ride A to D , ride F to B , ride F to E , ride D to E , ride D to C and ride B to C. Using the names of the rides , Draw a graph relating to their paths.

5) Zoom Zoom carnival has six crazy rides. There are only few paths to reach from one ride to another ride. Jade can go from ride A to B , ride A to F , ride E to C , ride F to B , ride F to E , ride D to E , ride D to C and ride B to C. Using the names of the rides , Draw a graph relating to their paths.

6) Sam's garden has six big apple trees named as A,B,C,D,E,F. Apple trees are connected to each other by a stone path. Sam can go from A to B , A to F , A to D , B to C , B to E , B to F , C to D , C to E , F to C , E to D and E to F. Using the names of the trees , Draw a graph relating to the connecting stone paths.

7) A bus leaves from Stop A to C , Stop A to D , Stop B to A , Stop D to B , Stop C to B and Stop D to C
Using the names of the stops , Draw a graph relating to the bus routes.

8) Island Oxygen has three bridges connecting to the city on the shore. There are roads connecting to the bridges A,B,C. The roads and bridges are connected as A to O , O to C , B to O , A to C , B to C , and B to A.
Using the names of the roads and bridges , Draw a graph relating to their paths.

9) Island Oxygen has three bridges connecting to the city on the shore. There are roads connecting to the bridges A,B,C. The roads and bridges are connected as A to O , O to C , B to O , A to C , and B to A. Using the names of the roads and bridges , Draw a graph relating to their paths.

10) Draw an undirected graph with the vertices A , B , C , D , E , F and edges as AB , CA , BC , DE , DF , FE , EA , CD , and BF.

11) Draw an undirected graph with the vertices A , B , C , D , E , F and edges as AD , AE , DE , DC , BE , FD , EF , CF , and BF.

12) Draw an undirected graph with the vertices A , B , C , D , E , F and edges as AB , AD , EA , DB , FB , BC , CA , CF , FE , and ED.

13) Draw an undirected graph with the vertices A , B , C , D , F and edges as AB , AD , DB , FB , BC , CF , and FD.

14) Draw an undirected graph with the vertices A , B , C , D , F and edges as AB , AD , DB , FB , BC , CA , CF , and FD.

15) Draw an undirected graph with the vertices A , B , C , D , E , F and edges as AB , AD , EA , DB , FB , BC , CF , FE , and ED.

16) Draw an undirected graph with the vertices A , B , C , D , E , F and edges as AB , AD , EA , FB , BC , CF , FE , and ED.

17) Draw an undirected graph with the vertices A , B , C , D , E and edges as AB , AD , EA , DB , BC , CE , and ED.

18) Draw an undirected graph with the vertices A , B , C , D , E and edges as AB , AD , EA , DB , BC , CA , CE , and ED.

19) Draw an undirected graph with the vertices A , B , C , D , E , F , G and edges as AB , AG, FG , EF , ED , CD , CB.

20) How many path of lengths equal to 2 exist in the below figure ? List all of them.

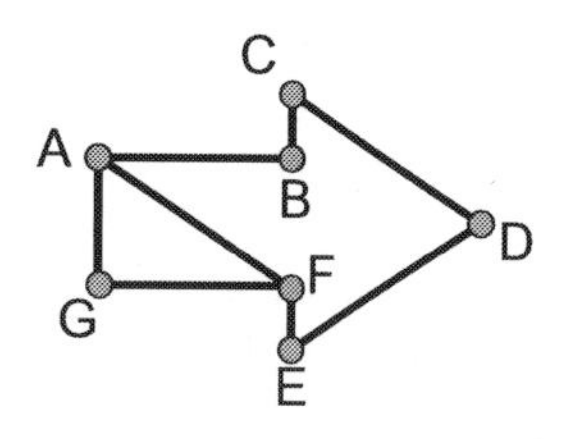

21) How many path of lengths equal to 2 exist in the below figure ? List all of them.

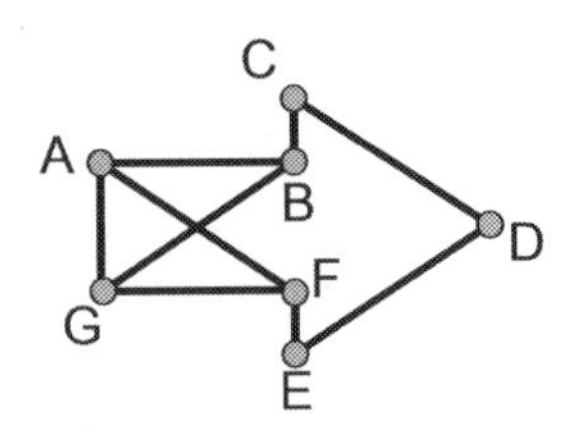

22) How many path of lengths equal to 2 exist in the below figure ? List all of them.

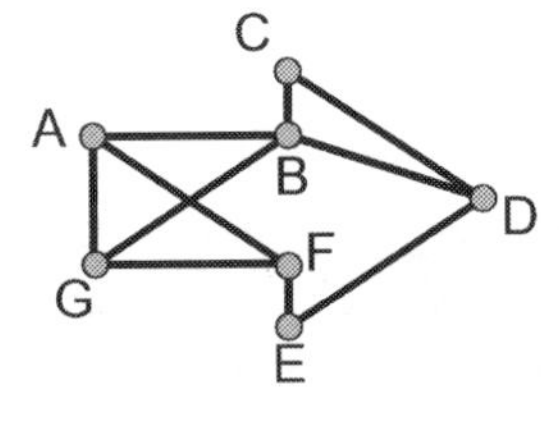

23) List all cycles of length 3 of the below figure.

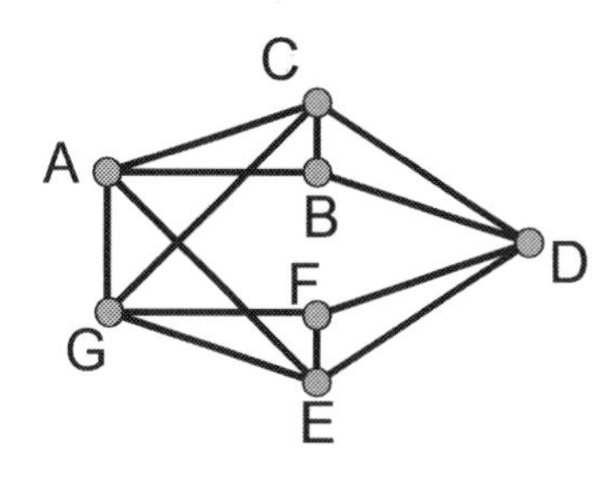

24) List all cycles of length 3 of the below figure.

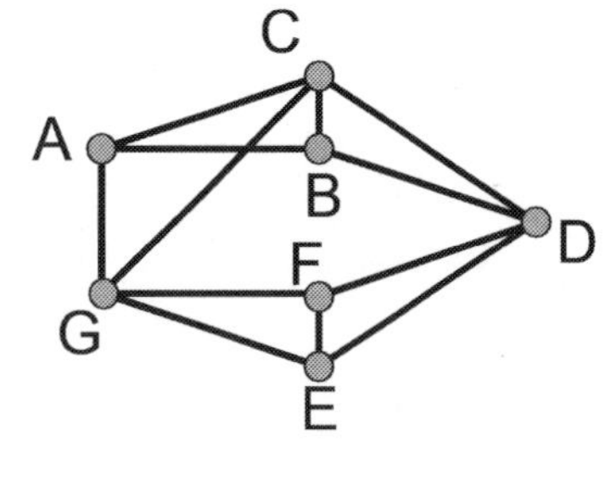

25) List all cycles of length 3 of the below figure.

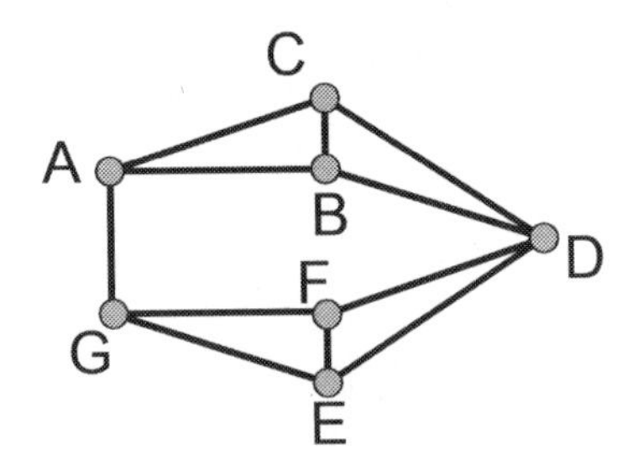

26) Find if the following graph is traversable or not ?

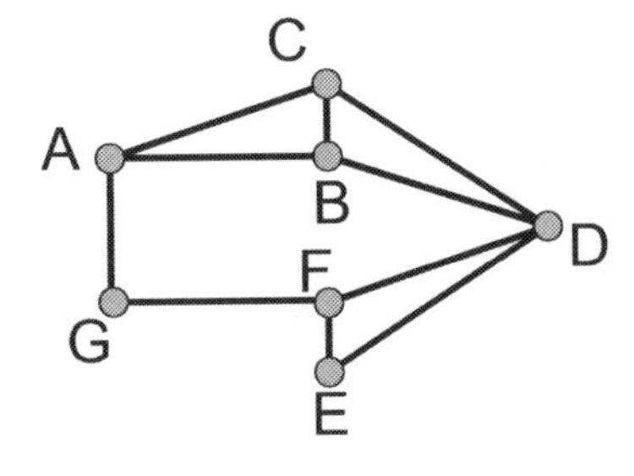

27) Find if the following graph is traversable or not ?

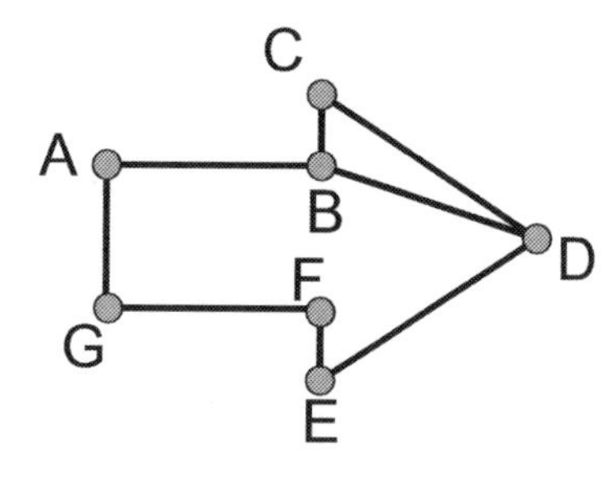

28) Find if the following graph is traversable or not ?

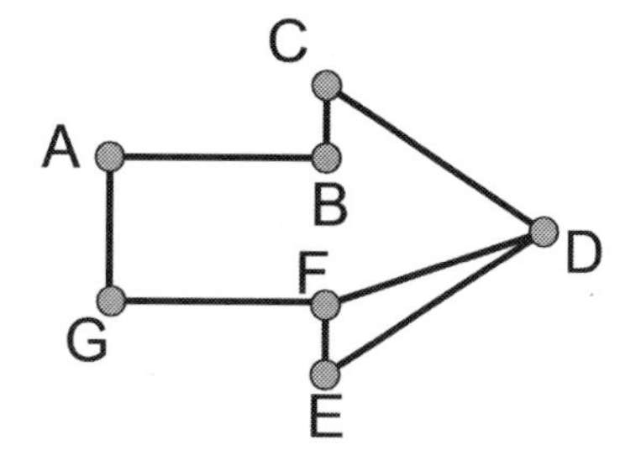

29) Find if the following graph is traversable or not ?

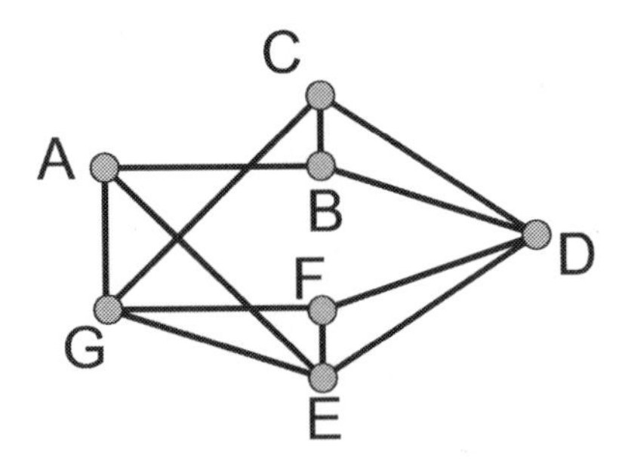

30) Find if the following graph is traversable or not ?

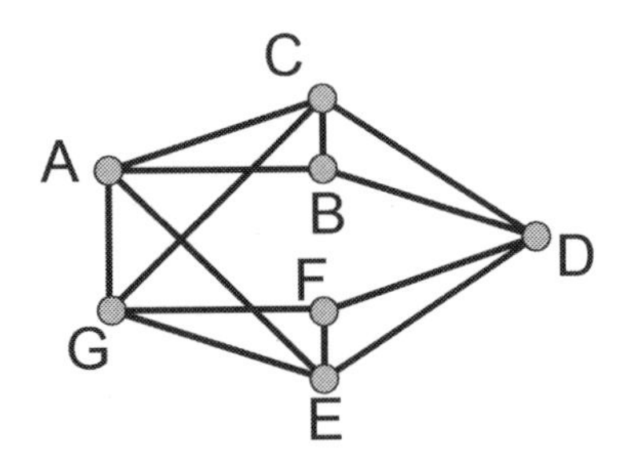

31) Find if the following graph is traversable or not ?

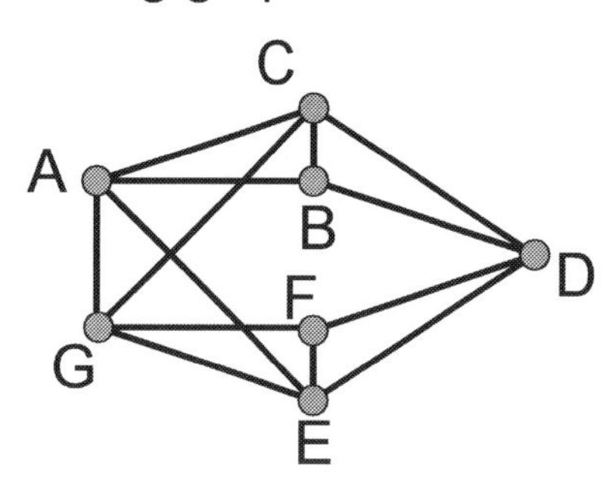

32) List all cycles of length 2 of the below figure.

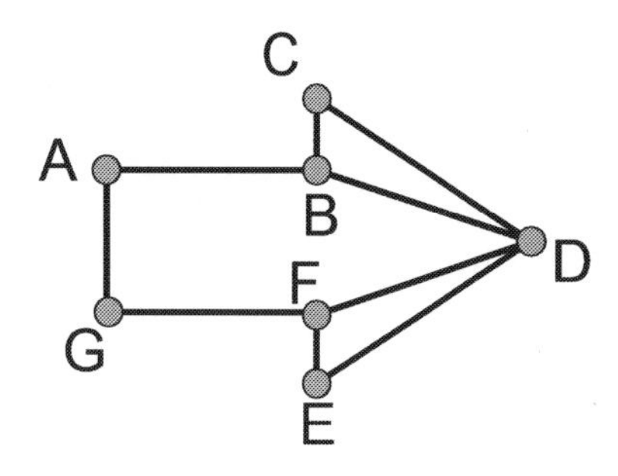

33) List all cycles of length 2 of the below figure.

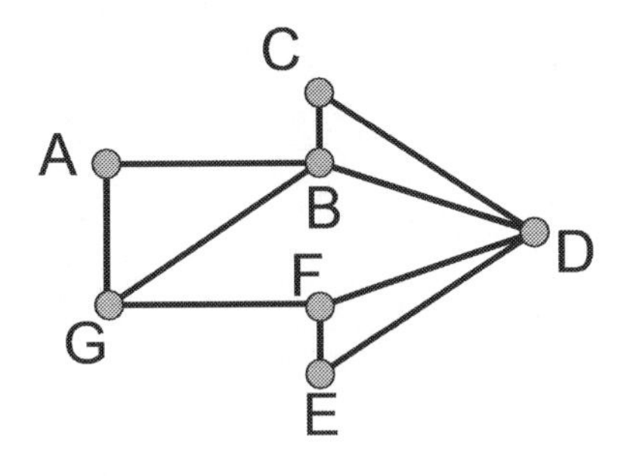

34) List all cycles of length 2 of the below figure.

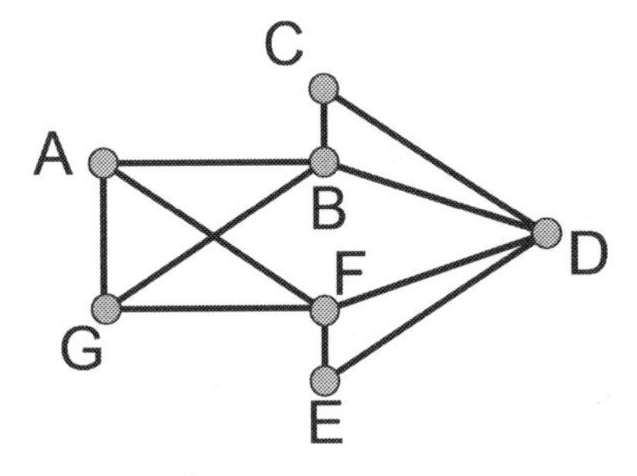

35) List all cycles of length 2 of the below figure.

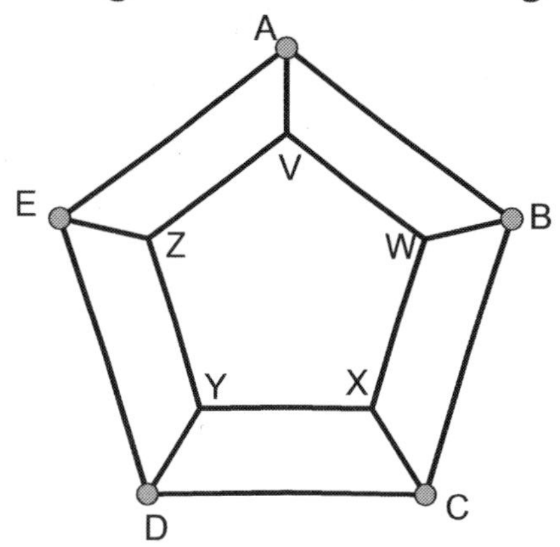

36) List all cycles of length 3 of the below figure.

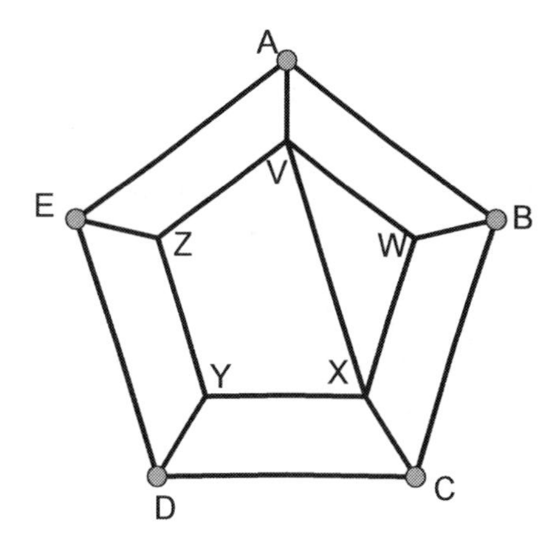

37) List all cycles of length 3 of the below figure.

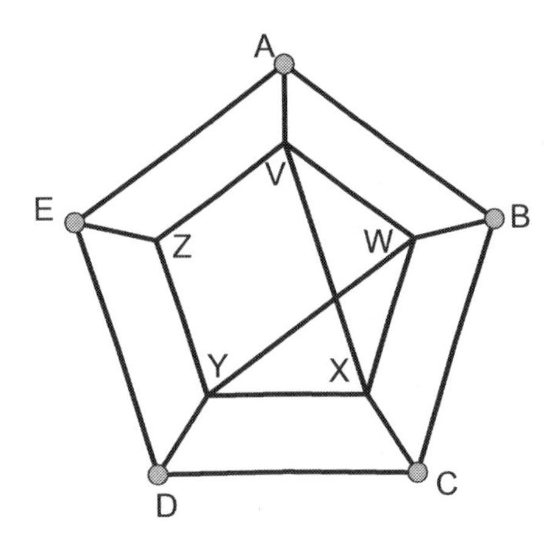

38) Find if the following graph is traversable or not ?

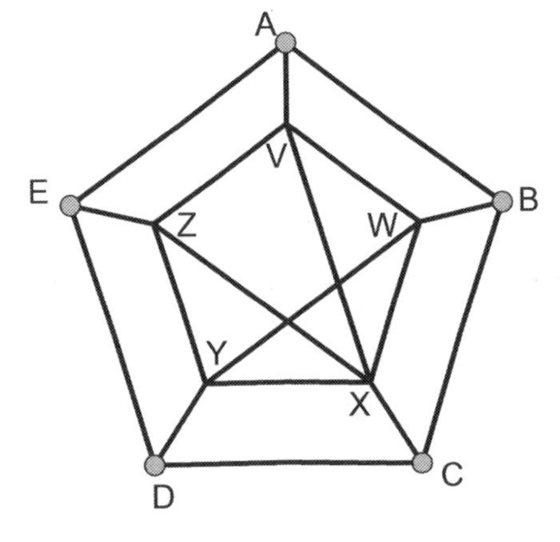

39) Find if the following graph is traversable or not ?

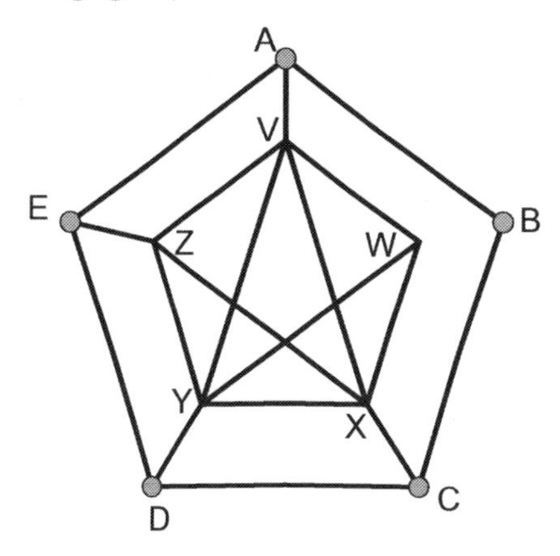

40) Find if the following graph is traversable or not ?

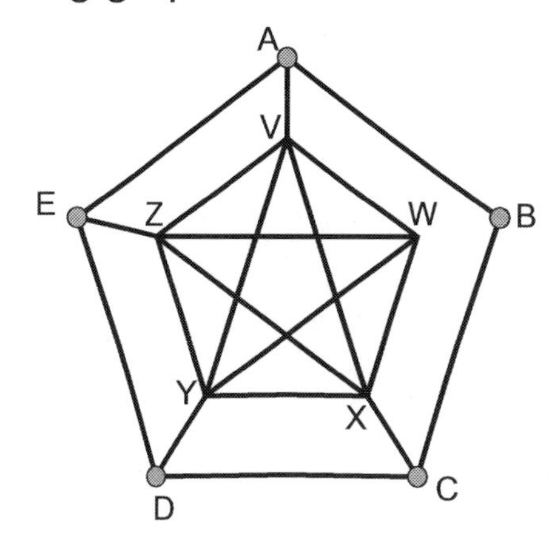

41) Find if the following graph is traversable or not ?

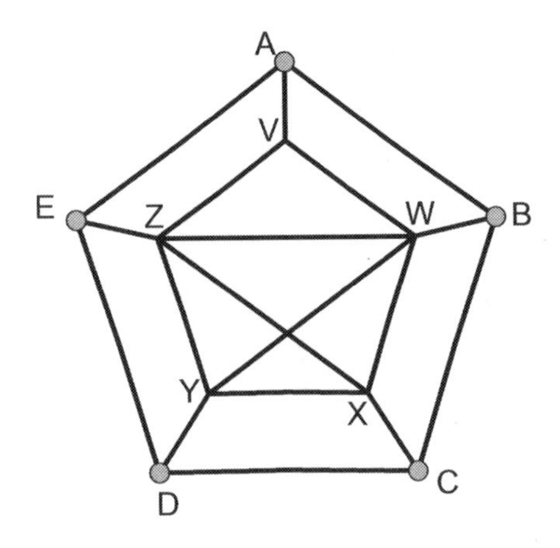

42) Find if the following graph is traversable or not ?

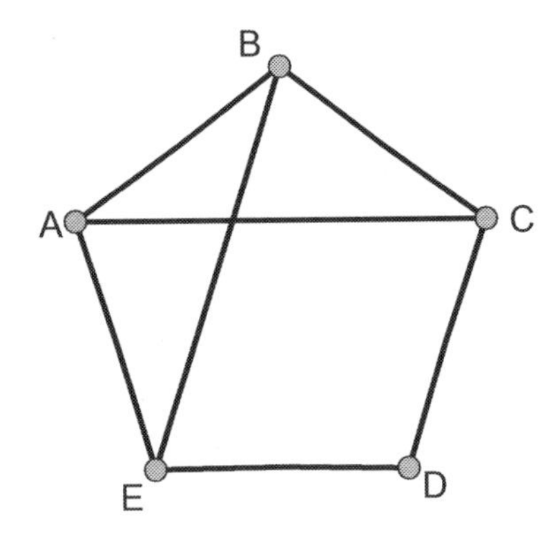

43) Find if the following graph is traversable or not ?

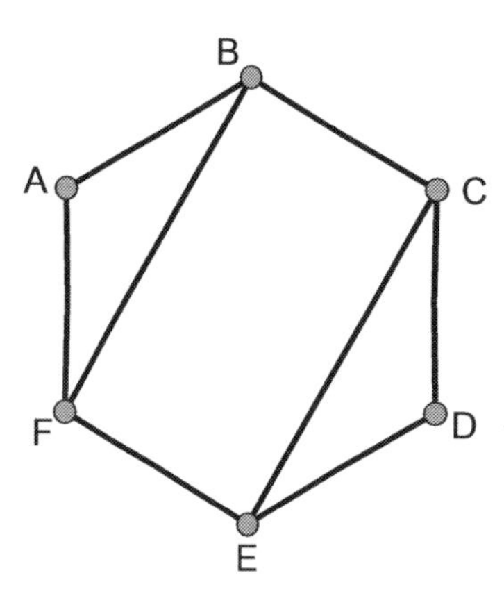

44) Find if the following graph is traversable or not ?

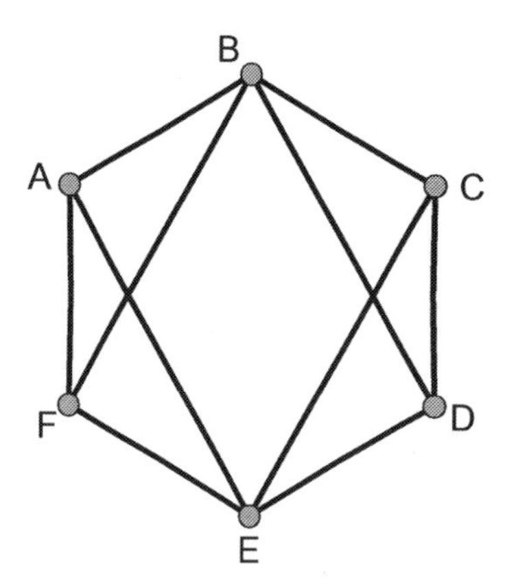

45) Find if the following graph is traversable or not ?

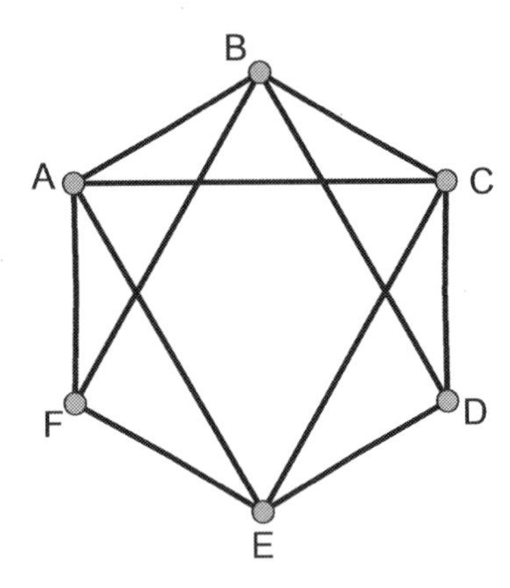

Answer Keys

MATH
KNOTS
Joy of learning...

1. Convert 120_8 to base 10.

   120_8 = 1*64 + 2*8 + 0*1 = 64 + 16 = 80_{10}

2. Convert 155_8 to base 10.

   155_8 = 1*64 + 5*8 + 5*1 = 64 + 40 + 5= 109_{10}

3. Convert 236_8 to base 10.

   236_8 = 2*64 + 3*8 + 6*1 = 128 + 24 + 6 = 158_{10}

4. Convert 334_8 to base 10.

   334_8 = 3*64 + 3*8 + 4*1 = 192 + 24 + 4 = 220_{10}

5. Convert 606_8 to base 10.

   606_8 = 6*64 + 0*8 + 6*1 = 384 + 0 + 6 = 390_{10}

6. Convert 328_8 to base 10.

   328_8 = 3*64 + 2*8 + 8*1 = 192 + 16 + 8 = 216_{10}

7. Convert 1121_8 to base 10.

   1121_8 = 1*512 + 1*64 + 2*8 + 1*1 = 512 + 64 + 16 + 1
   = 593_{10}

8. Convert 2156_8 to base 10.

   2156 = 2*512 + 1*64 + 5*8 + 6*1 = 1024 + 64 + 40 + 6
   = 1134_{10}

9. Convert 2177_8 to base 10.

   2177_8 = 2*512 + 1*64 + 7*8 + 7*1
   = 1024 + 64 + 56 + 7
   = 1151_{10}

10. Convert 32009_8 to base 10.

   32009_8 = 3*4096 + 2*512 + 0*64 + 0*8 + 9*1
   = 12288 + 1024 + 0 + 0 + 9
   = 13321_{10}

11. Convert 1079_8 to base 10.

1079_8 = 1*512 + 0*64 + 7*8 + 9*1
= 512 + 0 + 56 + 9
= 577_{10}

12. Convert $2AB_{16}$ to base 10.

$2AB_{16}$ = 2*256 + 10*16 + 11*1
= 512 + 160 + 11
= 683_{10}

13. Convert $3AD_{16}$ to base 10.

$3AD_{16}$ = 3*256 + 10*16 + 13*1
= 768 + 160 + 13
= 941_{10}

14. Convert $3CE_{16}$ to base 10.

$3CE_{16}$ = 3*256 + 12*16 + 14*1
= 768 + 192 + 14
= 974_{10}

15. Convert 2DE to base 16 to base 10.

$2DE_{16}$ = 2*256 + 13*16 + 14*1
= 512 + 208 + 14
= 734_{10}

16. Convert $1AF_{16}$ to base 10.

$1AF_{16}$ = 1*256 + 10*16 + 15*1
= 256 + 160 + 15 =
= 431_{10}

17. Convert $3DA_{16}$ to base 10.

$3DA_{16}$ = 3*256 + 13*16 + 10*1
= 768 + 208 + 10
= 986_{10}

18. Convert AAF_{16} to base 10. .

AAF_{16} = 10*256 + 10*16 + 15*1
= 2560 + 160 + 15
= 2735_{10}

19. Convert $A5E_{16}$ to base 10.

$A5E_{16}$ = 10*256 + 5*16 + 14*1
= 2560 + 80 + 14
= 2654_{10}

20. Convert $A3F_{16}$ to base 10. .

$A3F_{16}$ = 10*256 + 3*16 + 15*1
= 2560 + 48 + 15
= 2623_{10}

21. Convert $38C_{16}$ to base 10.

$38C_{16}$ = 3*256 + 8*16 + 12*1
= 768 + 128 + 12
= 908_{10}

22.Convert the following base 8 number to base 16:
24675_8

24675_8 = 010 100 110 111 101_2

= 010 1001 1011 1101_2
= $29BD_{16}$

23. Convert the following base 8 number to base 16:
44453_8

44453_8 = 100 100 100 101 011_2

=100 1001 0010 1011_2
= $492B_{16}$

24. Convert the following base 8 number to base 16:
31722_8

31722_8 = 011 001 111 010 010_2

=011 0011 1101 0010_2
= $33D2_{16}$

25. Convert the following base 8 number to base 16:

$$50116_8$$

$50116_8 = 101\ 000\ 001\ 001\ 110_2$

$=101\ 0000\ 0100\ 1110_2$

$= 504E_{16}$

26.Convert the following base 8 number to base 16:

$$25732_8$$

$25732_8 = 010\ 101\ 111\ 011\ 010_2$

$= 010\ 1011\ 1101\ 1010_2 = 2BDA_{16}$

27. Convert the following base 8 number to base 16:

$$204452_8$$

$204452_8 = 010\ 000\ 100\ 100\ 101\ 010_2$

$= 01\ 0000\ 1001\ 0010\ 1010_2$

$= 1092A_{16}$

28. Convert the following base 8 number to base 16:

$$150147_8$$

$150147_8 = 001\ 101\ 000\ 001\ 100\ 111_2$

$= 00\ 1101\ 0000\ 0110\ 0111_2$

$= 0D067_{16} = D067_{16}$

29. Convert the following base 8 number to base 16:

$$222206_8$$

$222206_8 = 010\ 010\ 010\ 010\ 000\ 110_2$

$= 01\ 0010\ 0100\ 1000\ 0110_2$

$= 12486_{16}$

30. Convert the following base 8 number to base 16:
303033_8

$303033_8 = 011\ 000\ 011\ 000\ 011\ 011_2$

$= 01\ 1000\ 0110\ 0001\ 1011_2$

$= 1861B_{16}$

31. Convert the following base 8 number to base 16:
127244_8

$127244_8 = 001\ 010\ 111\ 010\ 100\ 100_2$

$= 00\ 1010\ 1110\ 1010\ 0100_2$

$= 0AEA4_{16} = AEA4_{16}$

32. Convert $A52DE_{16}$ to Octal.

$A52DE_{16} = 1010\ 0101\ 0010\ 1101\ 1110_2$

$= 10\ 100\ 101\ 001\ 011\ 011\ 110_2$

$= 2451336_8$

33. Convert $DAC5A_{16}$ to Octal.

$DAC5A_{16} = 1101\ 1010\ 1100\ 0101\ 1010_2$

$= 11\ 011\ 010\ 110\ 001\ 011\ 010_2$

$= 3326132_8$

34. Convert $F785D_{16}$ to Octal

$F785D_{16} = 1111\ 0111\ 1000\ 0101\ 1101_2$

$= 11\ 110\ 111\ 100\ 001\ 011\ 101_2$

$= 3674135_8$

35. Convert $C980F_{16}$ to Octal

$C980F_{16} = 1100\ 1001\ 1000\ 0000\ 1111_2$

$= 11\ 001\ 001\ 100\ 000\ 001\ 111_2$

$= 3114017_8$

36. Convert $B13B7_{16}$ to Octal

$B13B7_{16} = 1011\ 0001\ 0011\ 1011\ 0111_2$

$= 10\ 110\ 001\ 001\ 110\ 110\ 111_2$

$= 2611667_8$

37. Convert $ED0B8_{16}$ to Octal

$ED0B8_{16} = 1110\ 1101\ 0000\ 1011\ 1000_2$

$= 11\ 101\ 101\ 000\ 010\ 111\ 000_2$

$= 3550270_8$

38. Convert $C45A2E_{16}$ to Octal

$C45A2E_{16} = 1100\ 0100\ 0101\ 1010\ 0010\ 1110_2$

$= 110\ 001\ 000\ 101\ 101\ 000\ 101\ 110_2$

$= 61055056_8$

39. Convert $AA5DE_{16}$ to Octal

$AA5DE_{16} = 1010\ 1010\ 0101\ 1101\ 1110_2$

$= 10\ 101\ 010\ 010\ 111\ 011\ 110_2$

$= 2522736_8$

40. Convert $CC13EE_{16}$ to Octal

$CC13EE_{16} = 1100\ 1100\ 0001\ 0011\ 1110\ 1110_2$

$= 110\ 011\ 000\ 001\ 001\ 111\ 101\ 110_2$

$= 63011756_8$

41. Convert $FEFE5F_{16}$ to Octal

$FEFE5F_{16} = 1111\ 1110\ 1111\ 1110\ 0101\ 1111_2$

$= 111\ 111\ 101\ 111\ 111\ 001\ 011\ 111_2$

$= 77577137_8$

42. How many more 1's are there in the binary representation of $5B4_{16}$ than in the binary representation of $C41_{16}$?

$C41_{16} = 1100\ 0100\ 0001_2$

$5B4_{16} = 0101\ 1011\ 0100_2$

There are 2 more 1's in $5B4_{16}$

43. How many more 1's are there in the binary representation of $7BE1_{16}$ than in the binary representation of $10D4_{16}$?

$7BE1_{16} = 0111\ 1011\ 1110\ 0001_2$

$10D4_{16} = 0001\ 0000\ 1101\ 0100_2$

There are 5 more 1's in $7BE1_{16}$

44. How many more 1's are there in the binary representation of $F4D8_{16}$ than in the binary representation of $18DE_{16}$?

$18DE_{16} = 0001\ 1000\ 1101\ 1110_2$

$F4D8_{16} = 1111\ 0100\ 1101\ 1000_2$

There is one more 1 in $F4D8_{16}$

45. How many more 1's are there in the binary representation of $CCD5_{16}$ than in the binary representation of $14AF_{16}$?

$CCD5_{16} = 1100\ 1100\ 1101\ 0101_2$

$14AF_{16} = 0001\ 0100\ 1010\ 1111_2$

There is one more 1 in $CCD5_{16}$

46. How many more 1's are there in the binary representation of $91FE_{16}$ than in the binary representation of $88C5_{16}$?

$88C5_{16} = 1000\ 1000\ 1100\ 0101_2$

$91FE_{16} = 1001\ 0001\ 1111\ 1110_2$

There are four more 1's in $91FE_{16}$

47. Express the sum in hexadecimal:

$54D1E_{16} + 1A2D_{16}$

$$\begin{array}{r} 54D1E_{16} \\ +\ \ 1AD8_{16} \\ \hline 567F6_{16} \end{array}$$

E + 8 = 14 + 8 = 22
22-16 = 6 (1 carry over)
1 + D + 1 = 1 + 13 + 1 = 15 = F
D + A = 13 + 10 = 23
23 - 16 = 7 (1 carry over)

48. Express the sum in hexadecimal:

$27E_{16} + 7CA_{16}$

$$\begin{array}{r} 27E_{16} \\ +\ 7CA_{16} \\ \hline A48_{16} \end{array}$$

E + A = 14 + 10 = 24
24 - 16 = 8 (1 carry over)

1 + C + 7 = 1 + 12 + 7 = 20
20 - 16 = 4 (1 carry over)

1 + 2 + 7 = 10 = A

49. Express the sum in hexadecimal:

$7AC_{16} + 8BD_{16}$

$$\begin{array}{r} 7AC_{16} \\ +\ 8BD_{16} \\ \hline 1069_{16} \end{array}$$

C + D = 12 + 13 = 25
25 - 16 = 9 (1 carry over)

1 + A + B = 1 + 10 + 11 = 22
22 - 16 = 6 (1 carry over)

1 + 8 + 7 = 16
16-16 = 0 (1 carry over)

50. Express the sum in hexadecimal:

$5EA_{16} + 6FC_{16}$

$$\begin{array}{r} 5EA_{16} \\ +\ 6FC_{16} \\ \hline CE6_{16} \end{array}$$

C + A = 12 + 10 = 22
22 - 16 = 6 (1 carry over)
1 + E + F = 1 + 14 + 15 = 30
30 - 16 = 14 = E (1 carry over)
1 + 5 + 6 = 12 = C

51. Express the sum in hexadecimal:

$68F_{16} + 8FB_{16}$

$$\begin{array}{r} 68F_{16} \\ +\ 8FB_{16} \\ \hline F8A_{16} \\ \hline \end{array}$$

F + B = 15 + 11 = 26
26 - 16 = 10 (1 carry over) = A

1 + 8 + F = 1 + 8 + 15 = 24
24 - 16 = 8 (1 carry over)

1 + 6 + 8 = 15 = F

52. Simplify the below in hexadecimal:

$B7F_{16} - 5DB_{16}$

$$\begin{array}{r} B7F_{16} \\ +\ 5DB_{16} \\ \hline 584_{16} \\ \hline \end{array}$$

F - B = 15 - 11 = 4
7 - D = 7 - 15 = 16 + 7 -15 (Bring 1 borrow ; 1 =16)
= 8
B - 5 -1 (DONT' FORGET The borrow given earlier)
= 11 - 5 - 1 = 5

53. Simplify the below in hexadecimal:

$DE2_{16} - 3EA_{16}$

$$\begin{array}{r} DE2_{16} \\ +\ 3\ FA_{16} \\ \hline 9E8_{16} \\ \hline \end{array}$$

2 - A = 2 - 10 = 16 + 2 - 10 (Bring 1 borrow ; 1 = 16)
= 8
E - F - 1(remember the borrow given)
= 14 - 15 - 1 = 14 -15 -1 + 16 (Bring 1 borrow ; 1 =16)
= 14 = E
D - 3 -1 (DONT' FORGET The borrow given earlier)
= 13 - 3 - 1 = 9

54. Simplify the below in hexadecimal:

$FF8_{16} - AAA_{16}$

$FF8_{16}$
$+ AAA_{16}$
$54E_{16}$

8 - A = 8 - 10 = 16 + 8 - 10 (Bring 1 borrow ; 1 = 16)
= 14 = E
F - A - 1(remember the borrow given)
= 15 - 10 - 1 = 15 -10 -1
= 4
F - A = 15 - 10 = 5

55. Simplify the below in hexadecimal:

$CFD_{16} - AB3_{16}$

CBD_{16}
$- AF3_{16}$
$1CA_{16}$

D - 3 = 13 - 3 = 10 = A

B - F (Bring 1 borrow ; 1 = 16)
= 11 - 15 + 16 = 12 = C
C - A = 12 - 10 - 1 (remember the borrow given)
= 1

56. Simplify the below in hexadecimal:

$DA9_{16} - 1DC_{16}$

$DA9_{16}$
$- 1DC_{16}$
ACD_{16}

9 - C = 9 - 12 (Bring 1 borrow ; 1 = 16)
= 9 - 12 + 16 = 13 = D
A - D (Bring 1 borrow ; 1 = 16)
= 10 - 13 - 1 (remember the borrow given) =
10 - 13 - 1 + 16 (Bring 1 borrow ; 1 = 16) = 12
= C
D - 1 = 12 - 1 - 1 (remember the borrow given)
= 10 = A

57. Express the sum in octa decimal:

$449_8 + 556_8$

$$\begin{array}{r} 449_8 \\ +556_8 \\ \hline 1227_8 \\ \hline \end{array}$$

9 + 6 = 15

15 - 8 = 7 (one carry over 1 = 8)

1 + 4 + 5 = 10

10 - 8 = 2 (one carry over 1 = 8)

4 + 5 + 1 = 10

10 - 8 = 2 (one carry over 1 = 8

58. Express the sum in octa decimal:

$5327_8 + 2117_8$

$$\begin{array}{r} 5327_8 \\ +2117_8 \\ \hline 7446_8 \\ \hline \end{array}$$

7 + 7 = 14

14 - 8 = 6 (one carry over 1 = 8)

1 + 2 + 1 = 4

3 + 1 = 4

5 + 2 = 7

59. Express the sum in octa decimal:

$4122_8 + 3588_8$

$$\begin{array}{r} 4122_8 \\ +3588_8 \\ \hline 7732_8 \\ \hline \end{array}$$

8 + 2 = 10

10 - 8 = 2 (one carry over 1 = 8)

1 + 2 + 8 = 11

11 - 8 = 3 (one carry over 1 = 8)

5 + 1 + 1 = 7

4 + 3 = 7

60. Simplify in octa decimal:

$6565_8 - 3985_8$

$$\begin{array}{r} 6565_8 \\ -\ 3985_8 \\ \hline 2360_8 \\ \hline \end{array}$$

6 - 8 = 6 - 8 + 8 (one borrow 1 = 8)
= 6

5 - 9 -1(remember one borrow was given ;1 = 8)
= 5 - 9 - 1 + 8 (one borrow 1 = 8
= 13 - 10 = 3
6 - 3 -1 (remember one borrow was given ;1 = 8)
= 6 - 4 = 2

61. Simplify in octa decimal:

$5904_8 - 1288_8$

$$\begin{array}{r} 5964_8 \\ -\ 1288_8 \\ \hline 4654_8 \\ \hline \end{array}$$

4 - 8 = 4 - 8 + 8 (one borrow 1 = 8)
= 4
6 - 8 -1 (remember one borrow was given ;1 = 8)
= 6 - 8 - 1 + 8 (one borrow 1 = 8
= 5
9 - 2 -1 (remember one borrow was given ;1 = 8)
= 9 - 3 = 6

62. Which is larger from the below, Order them from least to greatest

(A) 120_8 (B) 334_8 (C) 328_8

120_8= 1*64 + 2*8 + 0*1 = 64 + 16 = 80_{10}

334_8= 3*64 + 3*8 + 4*1 = 192 + 24 + 4 = 220_{10}

328_8 = 3*64 + 2*8 + 8*1 = 192 + 16 + 8 = 216_{10}

Largest is 334_8

Least to greatest 120_8 , 328_8, 334_8

63. Which is larger from the below, Order them from least to greatest

(A) $3AD_{16}$ (B) 236_8 (C) $2DE_{16}$

$3AD_{16}$ = 3*256 + 10*16 + 13*1
= 768 + 160 + 13
= 941_{10}

236_8 = 2*64 + 3*8 + 6*1 = 128 + 24 + 6 = 158_{10}

$2DE_{16}$ = 2*256 + 13*16 + 14*1
= 512 + 208 + 14
= 734_{10}

Largest is $3AD_{16}$

Least to greatest 236_8 , $2DE_{16}$, $3AD_{16}$

64. Which is larger from the below ,Order them from least to greatest

(A) 606_8 (B) $3DA_{16}$ (C) $1CE_{16}$

606_8 = 6*64 + 0*8 + 6*1 = 384 + 0 + 6 = 390_{10}

$3DA_{16}$ = 3*256 + 13*16 + 10*1
= 768 + 208 + 10
= 986_{10}

$1CE_{16}$ = 1*256 + 12*16 + 14*1
= 256 + 192 + 14
= 462_{16}

Largest is $3DA_{16}$

Least to greatest 606_8 , $1CE_{16}$, $3DA_{16}$

65. Which is larger from the below, Order them from least to greatest

(A) $2AB_{16}$ (B) 2156_8 (C) 1121_8

1121_8 = 1*512 + 1*64 + 2*8 + 1*1 = 512 + 64 + 16 + 6
= 598_{10}

$2AB_{16}$ = 2*256 + 10*16 + 11*1
= 512 + 160 + 11
= 683_{10}

2156_8 = 2*512 + 1*64 + 5*8 + 6*1 = 1024 + 64 + 40 + 6
= 1134_{10}
Largest is 2156_8
Greatest to least 2156_8 , $2AB_{16}$, 1121_8

66. Which is larger from the below, Order them from least to greatest

(A) 1079_8 (B) AAF_{16} (C) $38C_{16}$

1079_8 = 1*512 + 0*64 + 7*8 + 9*1
= 512 + 0 + 56 + 9
= 577_{10}

AAF_{16} = 10*256 + 10*16 + 15*1
= 2560 + 160 + 15
= 2735_{10}

$38C_{16}$ = 3*256 + 8*16 + 12*1
= 768 + 128 + 12
= 908_{10}

Largest is AAF_{16}

Greatest to least AAF_{16} , $38C_{16}$, 1079_8

67) What is the decimal value of the green component in the RGB code?

$6B8E23_{16}$

Green Component:

$8E_{16} = 8*16 + 14*1 = 128 + 14 = 142$

68) Is the decimal value of the blue and green component in the RGB code given below are equal is so what are the values?

$48D1CC_{16}$

Green Component: $D1_{16} = 13*16 + 1*1 = 209$

Blue Component: $CC_{16} = 12*16 + 12*1 = 204$

69) What are the decimal values of the Red, Green and Blue components in the RGB code?

$1E90FF_{16}$

Red Component: $1E_{16} = 1*16 + 14*1 = 30$

Green Component: $90_{16} = 9*16 + 0*1 = 144$

Blue Component: $FF_{16} = 15*16 + 15*1 = 255$

70) Is the decimal value of the blue and green component in the RGB code given below are equal is so what are the values, Which one is greater and by what value?

$F5DEB3_{16}$

Green Component: $DE_{16} = 13*16 + 14*1 = 222$

Blue Component: $B3_{16} = 11*16 + 3*1 = 179$

Green component is more by 43

71) Is the decimal value of the blue and green component in the RGB code given below are equal is so what are the values, Which one is greater and by what value?

$A0522D_{16}$

Red Component: $A0_{16} = 10*16 + 10*1 = 160$

Green Component: $52_{16} = 7*16 + 0*1 = 82$

Red component is more by 78

1.Evaluate the following postfix expression. Note that all numbers are not single digit.

25 25 / 72*8 − +

SOLUTIONS:

= (25 / 25) ((7 * 2) 8 −) +
= 1 (14 − 8) +
= 1 6 + = 1 + 6
= 7

2. Evaluate the following postfix expression. Note that all numbers are single digit.

6 5 * 3 3 + / + 3 3 * 2 -

SOLUTIONS:

(6 * 5) / (3 + 3) + 3 * 3 − 2
= (30 / 6) + 9 - 2 = 5 + 9 - 2
= 12

3. Evaluate the following postfix expression. Note that all numbers are not single digit.

15 3 / 5 9 * 5 − +

SOLUTIONS:

= (15 / 3) ((5 * 9) 5 −) +
= 5 (45 − 5) +
= 5 40 + = 5 + 40
= 45

4. Evaluate the following postfix expression. Note that all numbers are not single digit.

88 8 /8 8 * 8 − +

SOLUTIONS:

= (88 / 8) ((8 * 8) 8 −) +
= 11 (64 − 8) +
= 11 56 + = 11 + 56
= 67

5. Evaluate the following postfix expression. Note that all numbers are not single digit.

65 5 / 7 2 * 7 - +

SOLUTIONS:

= (6 5 / 5) ((7 * 2) 7 −) +
= 13 (14 − 7) +
= 13 7 + = 13 + 7
= 20

6. Evaluate the following postfix expression. Note that some numbers are not single digit.

9 4 * 9 9 + / 10 4 * + 5 -

SOLUTIONS:

= (9 * 4) / (9 + 9) + 10 * 4 − 5
= (36 / 18) + 40 - 5 = 2 + 35 = 37

7. Evaluate the following postfix expression if all numbers are positive

6 3 4 + 1 ^ * 12 2 / 4 * −

SOLUTIONS:

= 6 (3 + 4) 1 ^ * (12 / 2) 4 * −
= 6 7 1 ^ * 6 4 * −
= 6 (7 ^ 1) * (6 * 4) −
= 6 7 * 24 − = (6 * 7) 24 −
= 42 24 − = 42 − 24 = 18

8. Evaluate the following postfix expression if all numbers are single digits.

7 1 1 + 2 ^ * 14 7 / 7 * −

SOLUTIONS:

= 7 (1 + 1) 2 ^ * (14 / 7) 7 * −
= 7 2 2 ^ * 2 7 * −
= 7 (2 ^ 2) * (2 * 7) −
= 7 4 * 14 − = (7 * 4) 14 −
= 28 14 − = 28 − 14
= 14

9. Evaluate the following postfix expression if all numbers are single digits.

2 2 2 + 2 ^ * 2 2 / 2 * −

SOLUTIONS:

= 2 (2 + 2) 2 ^ * (2 / 2) 2 * −
= 2 4 2 ^ * 1 2 * −
= 2 (4 ^ 2) * (1 * 2) −
= 2 16 * 2 − = (2 * 16) 2 −
= 32 2 − = 32 − 2 = 30

10. Evaluate the following postfix expression if all numbers are positive

8 4 8 + 2 ^ * 32 8 / 4 * −

SOLUTIONS:

= 8 (4 + 8) 2 ^ * (32 / 8) 4 * −
= 8 12 2 ^ * 4 4 * −
= 8 (12 ^ 2) * (4 * 4) −
= 8 144 * 16 − = (8 * 144) 16 −
= 1152 16 − = 1152 − 16
= 1136

11. Evaluate the following postfix expression if all numbers are not single digits.

10 6 2 - 3 ^ * 50 5 / 4 * +

SOLUTIONS:

= 10 (6 - 2) 3 ^ * (50 / 5) 4 * +
= 10 4 3 ^ * 10 4 * +
= 10 (4 ^ 3) * (10 * 4) +
= 10 64 * 40 + = (10 * 64) 40 +
= 640 40 + = 640 + 40
= 680

12. Evaluate the following prefix expression. Note that all numbers are single digits in the expression.

    + / + 5 5 5 ^ 5 1

SOLUTIONS:

= + / (5 + 5) 5 (5 ^ 1)
= + / 10 5 5 = + (10 / 5) 5
= + 2 5 = 5 + 2
= 7

13. Evaluate the following prefix expression if all numbers are not single digits.

    − * + 22 11 3 * 7 / 15 5

SOLUTIONS:

= − * (22 + 11) 3 * 7 (15 / 5)
= − * 33 3 * 7 3 = − (33 * 3)(7 * 3)
= − 99 21 = 99 − 21 = 78

14. Evaluate the following prefix expression. Note that all numbers are single digits in the expression.

    − * + 6 4 6 ^ 3 1

SOLUTIONS:

= − * (6 + 4) 6 (3 ^ 1)
= − * 10 6 3 = - 10 * 6 3 = − 60 3
= 60 - 3
= 57

15. Evaluate the following prefix expression if all numbers are single digits.

    * * + 1 1 1 * 1 / 4 2

SOLUTIONS:

= * * (1 + 1) 1 * 1 (4 / 2)
= * * 2 1 * 1 2 = *(2 * 1) (1 * 2)
= * 2 2 = 2 * 2 = 4

16. Evaluate the following prefix expression. Note that all numbers are single digits in the expression.

- * + 5 6 5 ^ 5 1

SOLUTIONS:

= −* (5 + 6) 5 (5 ^ 1)
= −* 11 5 5 = -11 * 5 5 = -55 5
= 55 − 5
= 50

17. Evaluate the following prefix expression if all numbers are single digits.

+ * - 15 4 5 + 4 / 8 8

SOLUTIONS:

= + * (15 - 4) 5 + 4 (8 / 8)
= + * 11 5 + 4 1 = - (11 * 5) (4 + 5)
= + 55 9 = 55 + 9 = 64

18. Evaluate the following prefix expression. Note that all numbers are single digits in the expression.

* ^ / 2 2 2 + 2 2

SOLUTIONS:

= * ^ (2 / 2) 2 (2 + 2)
= * ^ 1 2 4 = * 1 ^ 2 4 = * 1 4
= 1 * 4
= 4

19. Evaluate the following prefix expression if all numbers are single digits.

+ ^ -7 0 2 * 4 / 6 2

SOLUTIONS:

= + ^ (7 - 0) 2 * 4 (6 / 2)
= + ^ 7 2 * 4 3 = + (7 ^ 2) (4 * 3)
= + 49 12 = 49 + 12 = 61

20. Evaluate the following prefix expression. Note that all numbers are single digits in the expression.

^ * / + 4 4 4 ^ 4 2 0

SOLUTIONS:
= ^ * / (4 + 4) 4 (4 ^ 2) 0
= ^ * / 8 4 16 0 = ^ * (8 / 4) 16 0
= ^ * 2 16 0 = ^ 2 * 16 0
= ^ 32 0 = 32 ^ 0 = 1

21. Evaluate the following prefix expression if all numbers are single digits.

+ * + 8 2 8 * 4 / 8 2

SOLUTIONS:

= + * (8 + 2) 8 * 4 (8 / 2)
= + * 10 8 * 4 4 = + (10 * 8) (4 * 4)
= + 80 16 = 80 + 16 = 96

22. Evaluate the following prefix expression if all numbers are not single digits.

+ * ^ 99 0 5 ^ 6 / 4 2

SOLUTIONS:

= + * (99 ^ 0) 5 ^ 6 (4 / 2)
= + * 1 5 ^ 6 2 = + (1 * 5) (6 ^ 2)
= + 5 36 = 5 + 36 = 41

23.Convert the infix expression to a postfix expression. Perform all operations in the same order as the given expression.

5 * 4 / 1 − 3 ^ 3 + 7 * 2 − 4 ^ 2

SOLUTION:

5 * 4 / 1 − 3 ^ 3 + 7 * 2 − 4 ^ 2
= ((5 4 *) / 1) − (3 3 ^) + (7 2 *) − 4 2 ^
= (5 4 * 1 /) − (3 3 ^) + (7 2 *) − 4 2 ^
= ((5 4 * 1 /) (3 3 ^) −) + (7 2 *) − 4 2 ^
= ((5 4 * 1 /) (3 3 ^) −) (7 2 *) + 4 2 ^ −
= 5 4 * 1 / 3 3 ^ − 7 2 * + 4 2 ^ −

24. Convert the infix expression to a postfix expression. Perform all operations in the same order as the given expression.

(9 − 5) / 1 * 5 / (9 − 3)

SOLUTION:

(9 − 5) / 1 * 5 / (9 − 3)
= (((9 5 −) / 1) * 5) / (9 3 −)
= ((9 5 − 1 /) * 5) / (9 3 −)
= ((9 5 − 1 /) 5 *) / (9 3 −)
= ((9 5 − 1 /) 5 *) (9 3 −) /
= 9 5 − 1 / 5 * 9 3 − /

25. Convert the infix expression to a postfix expression. Perform all operations in the same order as the given expression.

7 − 8 * 5 + 4 * 4 − 4

SOLUTION:

7 − 8 * 5 + 4 * 4 − 2
= 7 − (8 * 5) + (4 * 4) − 2
= 7 − (8 5 *) + (4 4 *) − 2
= (7 (8 5 *) −) + ((4 4 *) 2 −)
= ((7 (8 5 *) −) (4 4 *) 2 −) +
= 7 8 5 * − 4 4 * 2 − +

26. Convert the infix expression to a postfix expression. Perform all operations in the same order as the given expression.

$(6 + 4)^3 * (9 - 5)^2 / 4$

SOLUTION:

(6 + 4)^3 * (9 – 5)^2 / 4
= (+ 6 4)^3 * (– 9 5)^2 / 4
= (^ (+ 6 4) 3) * (^ (– 9 5) 2) / 4
= (* (^ (+ 6 4) 3) (^ (– 9 5) 2)) / 4
= / (* (^ (+ 6 4) 3) (^ (– 9 5) 2)) 4
= / * ^ + 6 4 3 ^ – 9 5 2 4

27. Convert the infix expression to a prefix expression. Perform all operations in the same order as the given expression.

7 + 5 * 2 – 5 * 3 + 9

SOLUTION:

7 + 5 * 2 – 5 * 3 + 9
= 7 + (5 * 2) –(5 * 3) + 9
= 7 + (* 5 2) – (* 5 3) + 9
= (+ 7 (* 5 2)) – (* 5 3) + 9
= (– (+ 7 (* 5 2)) (* 5 3)) + 9
= + (– (+ 7 (* 5 2)) (* 5 3)) 9
= + – + 7 * 5 2 * 5 3 9

28. Convert the infix expression to prefix. Perform all operations in the same order as the expression.

9 / (6 – 3) + (4 + 5)^3 – 5 * 7

SOLUTION:

9 / (6 – 3) + (4 + 5) ^ 3 – 5 * 7
= 9 / (– 6 3) + (+ 4 5) ^ 3 – * 5 7
= (/ 9 (– 6 3)) + (^ (+ 4 5) 3) – * 5 7
= + (/ 9 (– 6 3)) (^ (+ 4 5) 3) – * 5 7
= - + / 9 – 6 3 ^ + 4 5 3 – * 5 7

29. Convert the infix expression to a prefix expression. Perform all operations in the same order as the given expression.

$5^{(2+3)} * (9 - 1)^3$

SOLUTION:

5^(2 + 3) * (9 − 1) ^ 3
= 5 ^ (+ 2 3) * (− 9 1) ^ 3
= ^ 5 (+ 2 3) * (− 9 1) ^ 3
= * ^ 5 (+ 2 3)(^ (− 9 1) 3)
= * ^ 5 + 2 3 ^ − 9 1 3

30. Convert the infix expression to a prefix expression. Perform all operations in the same order as the given expression.

$3^{(3+3)} / (7 - 4)^3$

SOLUTION:

3^(3 + 3) / (7 − 4) ^ 3
= 3 ^ (+ 3 3) / (− 7 4) ^ 3
= ^ 3 (+ 3 3) / (− 7 4) ^ 3
= / ^ 3 (+ 3 3)(^ (− 7 4) 3)
= / ^ 3 + 3 3 ^ − 7 4 3

31. Translate the following postfix expression to prefix if all numbers are single digits.

9 8 + 7 * 9 7 8 + * −

SOLUTION:

9 8 + 7 * 9 7 8 + * −

= (9 + 8) 7 * 9 (7 + 8) * −
= ((9 + 8) * 7) (9 * (7 + 8))
= ((9 + 8) * 7) − (9 * (7 + 8))
= ((+ 9 8) * 7) − (9 * (+ 7 8))
= (* (+ 9 8) 7) − (* 9 (+ 7 8))
= − (* (+ 9 8) 7) (* 9 (+ 7 8))
= − * + 9 8 7 * 9 + 7 8

32. Translate the following postfix expression to prefix if all numbers are single digits.

4 3 − 5 4 + 3 * 5 3 4 + * −

SOLUTION:

4 3 − 5 4 + 3 * 5 3 4 + * −

= (4 − 3) (5 + 4) 3 * 5 (3 + 4) * −
= (4 − 3) ((5 + 4) * 3) (5 * (3 + 4))
=(4 − 3) ((5 + 4) * 3) − (5 * (3 + 4))
= (− 4 3) ((+ 5 4) * 3) − (5 * (+ 3 4))
= (* (− 4 3) (+ 5 4) 3) − (* 5 (+ 3 4))
= − (* (− 4 3) (+ 5 4) 3) (* 5 (+ 3 4))
= − * − 4 3 + 5 4 3 * 5 + 3 4

33. Translate the following postfix expression to prefix if all numbers are single digits.

6 9 + 3 / 6 3 1 + * −

SOLUTION:

6 9 + 3 / 6 3 1 + * −

= (6 + 9) 3 / 6 (3 + 1) * −
= ((6+ 9) / 3) (6 * (3 + 1)) −
= ((6 + 9) / 3) − (6 * (3 + 1))
= ((+ 6 9) / 3) − (6 * (+ 3 1))
= (/ (+ 6 9) 3) − (* 6 (+ 3 1))
= − (/ (+ 6 9) 3) (* 6 (+ 3 1))
= − / + 6 9 3 * 6 + 3 1

34. Translate the following postfix expression to prefix if all numbers are single digits.

7 7 + 7 / 1 7 7 + * −

SOLUTION:

7 7 + 7 / 1 7 7 + * −

= (7 + 7) 7 / 1 (7 + 7) / −
= ((7 + 7) / 7) (1 / (7 + 7)) −
= ((7 + 7) / 7) − (1 / (7 + 7))
= ((+ 7 7) / 7) − (1 (+ 7 7))
= (/ (+ 7 7) 7) − (/ 1 (+ 7 7))
= − (/ (+ 7 7) 7) (/ 1 (+ 7 7))
= − / + 7 7 7 / 1 + 7 7

35. Translate the following prefix expression to postfix. Note that all numbers are single digits in the expression.

* ^ + 7 4 9 – 8 1

SOLUTION:

* ^ + 7 4 9 – 8 1
= * ^ (7 + 4) 9 (8 – 1)
= * ((7 + 4) ^ 9) (8 – 1)
= ((7 + 4) ^ 9) * (8 – 1)
= (((7 4 +) 9 ^) (8 1 –) *)
= 7 4 + 9 ^ 8 1 – *

36. Translate the following prefix expression to postfix. Note that all numbers are single digits in the expression.

/ ^ + 1 2 5 – 5 2

SOLUTION:

/ ^ + 1 2 5 – 5 2
= / ^ (1 + 2) 5 (5 – 2)
= / ((1 + 2) ^ 5) (5 – 2)
= ((1 + 2) ^ 5) / (5 – 2)
= (((1 2 +) 5 ^) (5 2 –) /)
= 1 2 + 5 ^ 5 2 – /

37. Translate the following prefix expression to postfix. Note that all numbers are single digits.

* / – 6 2 4 * + 3 1 + 5 3

SOLUTION:

* / – 6 2 4 * + 3 1 + 5 3
= * / (6 – 2) 4 * (3 + 1) (5 + 3)
= * ((6 – 2) / 4) ((3 + 1) * (5 + 3))
= ((6 – 2) / 4) * ((3 + 1) * (5 + 3))
= ((6 2 –) 4 /) * ((3 1 +) (5 3 +) *)
= ((6 2 –) 4 /) ((3 1 +) (5 3 +) *) *
= 6 2 – 4 / 3 1 + 5 3 + * *

38. Translate the following prefix expression to postfix. Note that all numbers are single digits.

* / + 5 5 1 * – 5 1 * 1 3

SOLUTION:

* / + 5 5 1 * – 5 1 * 1 3
= * / (5 + 5) 1 * (5 – 1) (1 * 3)
= * ((5 + 5) / 1) ((5 – 1) * (1 * 3))
= ((5 + 5) / 1) * ((5 – 1) * (1 * 3))
= ((5 5 +) 1 /) * ((5 1 –) (1 3 *) *)
= ((5 5 +) 1 /) ((5 1 –) (1 3 *) *) *
= 5 5 + 1 / 5 1 – 1 3 * * *

39. Translate the following prefix expression to postfix. Note that all numbers are single digits.

+ / – 6 1 6 * + 1 6 1

SOLUTION:

+ / – 6 1 6 * + 1 6 1
= + / (6 – 1) 6 * (1 + 6) 1
= + ((6 – 1) / 6) ((1 + 6) * 1)
= ((6 – 1) / 6) + ((1 + 6) * 1)
= ((6 1 –) 6 /) ((1 6 +) 1 *) +
= 6 1 – 6 / 1 6 + 1 * +

40. Translate the following prefix expression to postfix. Note that All numbers are single digits.

+ / – 4 3 4 * + 3 4 3

SOLUTION:

+ / – 4 3 4 * + 3 4 3
= + / (4 – 3) 4 * (3 + 4) 3
= + ((4 – 3) / 4) ((3 + 4) * 3)
= ((4 – 3) / 4) + ((3 + 4) * 3)
= ((4 3 –) 4 /) ((3 4 +) 3 *) +
= 4 3 – 4 / 3 4 + 3 * +

41. Translate the following prefix expression to postfix. Note that all numbers are single digits.

+ / + 4 7 6 * * 7 6 4

SOLUTION:

+ / + 4 7 6 * * 7 6 4
= + / (4 + 7) 6 * (7 * 6) 4
= + ((4 + 7) / 6) ((7 * 6) * 4)
= ((4 + 7) / 6) + ((7 * 6) * 4)
= ((4 7 +) 6 /) ((7 6 *) 4 *) +
= 4 7 + 6 / 7 6 * 4 * +

42. Translate the following prefix expression to postfix. Note that all numbers are single digits.

+ / − 7 3 7 * * 7 3 7

SOLUTION:

+ / − 7 3 7 * * 7 3 7
= + / (7 − 3) 7 * (7 * 3) 7
= + ((7 − 3) / 7) ((7 * 3) * 7)
= ((7 − 3) / 7) + ((7 * 3) * 7)
= ((7 3 −) 7 /) ((7 3 *) 7 *) +
= 7 3 − 7 / 7 3 * 7 * +

43. Translate the following prefix expression to postfix. Note that all numbers are single digits.

− / − 2 1 7 * * 2 2 1

SOLUTION:

− / − 2 1 7 * * 2 2 1
= − / (2 − 1) 7 * (2 * 2) 1
= − ((2 − 1) / 7) ((2 * 2) * 1)
= ((2 − 1) / 7) − ((2 * 2) * 1)
= ((2 1 −) 7 /) ((2 2 *) 1 *) −
= 2 1 − 7 / 2 2 * 1 * −

44. Translate the follow postfix expression to prefix. Note that all numbers are single digits.

3 3 * 2 2 + − 3 2 * 2 − +

SOLUTION:

3 3 * 2 2 + − 3 2 * 2 − +
= (3 * 3) (2 + 2) − (3 * 2) 2 − +
= ((3 * 3) − (2 + 2)) ((3 * 2) − 2) +
= ((3 * 3) − (2 + 2)) + ((3 * 2) − 2)
= ((* 3 3) − (+ 2 2)) + ((* 3 2) − 2)
= (− (* 3 3) (+ 2 2)) + (− (* 3 2) 2)
= + (− (* 3 3) (+ 2 2)) (− (* 3 2) 2)
= + − * 3 3 + 2 2 − * 3 2

45. Translate the follow postfix expression to prefix. Note that all numbers are single digits.

9 1 * 5 2 + − 6 5 * 6 − −

SOLUTION:

9 1 * 5 2 + − 6 5 * 6 − −
= (9 * 1) (5 + 2) − (6 * 5) 6 − −
= ((9 * 1) − (5 + 2)) ((6 * 5) − 6) −
= ((9 * 1) − (5 + 2)) − ((6 * 5) − 6)
= ((* 9 1) − (+ 5 2)) − ((* 6 5) − 6)
= (− (* 9 1) (+ 5 2)) − (− (* 6 5) 6)
= − (− (* 9 1) (+ 5 2)) (− (* 6 5) 6)
= − − * 9 1 + 5 2 − * 6 5 6

Evaluate the following expression as either TRUE or FALSE:

1.

(((4 + 5) <= 11) AND ((9 - 5)>= 3 * 2)) OR NOT (7 * 7 < 5 * 5)
((TRUE) AND (FALSE)) OR TRUE = FALSE OR TRUE = TRUE

2.

((22/2) >= 22) AND (2 * 5)>= 2 * 2) OR NOT (3 + 7 >= 2 * 5)
(FALSE AND TRUE) OR NOT(TRUE) = FALSE OR FALSE = FALSE

3.

((10 - 1) <= (10 + 1) AND ((3 * 9)>= 3 * 6)) OR NOT (10 * 7 > 6 * 9)
(TRUE AND TRUE) OR FALSE = TRUE OR FALSE = TRUE

4.

((7 + 3) <= 4 + 5) AND ((6 + 1)>= 4 * 1) OR NOT (99/11 > = 3 * 3)
(FALSE AND TRUE) OR NOT(TRUE) = FALSE OR FALSE = FALSE

5.

((1/5 <= 1/4) AND ((2/3) >= 2 * (1/3))) OR NOT (13 + 5 = 5 + 13)
(TRUE AND TRUE) OR FALSE = TRUE OR FALSE = TRUE

6.

NOT (6^2 < 4 * 6) OR NOT (10/ 5 ≥ 3 AND 12 * 4 > 10 * 5)
FALSE OR NOT (FALSE AND FALSE) = FALSE OR NOT(FALSE) = FALSE OR TRUE = TRUE

7.

NOT (3^3 < 5^2) OR (10/ 5 ≥ 3 AND 12 * 4 > 10 * 5)
TRUE OR (FALSE AND FALSE) = TRUE OR FALSE = TRUE

8.

NOT (19-12 < = 3) OR (64 / 8 ≥ 4 * 2 AND 6 * 3 > 3 * 1)
TRUE OR (TRUE AND TRUE) = TRUE OR TRUE = TRUE

9.

NOT (10^2 < 2 ^4) OR (72 / 6 ≥ 12 AND 40 * 2 >9 * 8)
TRUE OR (TRUE AND TRUE) = TRUE OR TRUE = TRUE

10.

NOT (50 + 50 < = 90 + 10) OR (49 / 7 ≥ 9 * 6 AND 8 * 6 >5 * 6)
FALSE OR (FALSE AND TRUE) = FALSE OR FALSE = FALSE

11.
How many ordered pairs make the following statement FALSE?

NOT A OR (A AND (A AND B))

A	B	NOT A	A AND B	A AND (A AND B)	NOT A OR (A AND (A AND B))
1	1	0	1	1	1
0	1	1	0	0	1
1	0	0	0	0	0
0	0	1	0	0	1

12.

(A OR B) AND (A OR NOT B)

A	B	NOT B	A OR B	A OR NOT B	(A OR B) AND (A OR NOT B)
1	1	0	1	1	1
0	1	0	1	0	0
1	0	1	1	1	1
0	0	1	0	1	0

13.

A OR NOT (A AND B)

A	B	A AND B	NOT (A AND B)	A OR NOT (A AND B)
1	1	1	0	1
0	1	0	1	1
1	0	0	1	1
0	0	0	1	1

14.

A AND NOT (A OR B)

A	B	A OR B	NOT (A OR B)	A AND NOT (A OR B)
1	1	1	0	0
0	1	1	0	0
1	0	1	0	0
0	0	0	1	0

15.

NOT A AND (A AND(A OR B))

A	B	NOT A	A OR B	A AND (A OR B)	NOT A AND (A AND (A OR B))
1	1	0	1	1	0
0	1	1	1	0	0
1	0	0	1	1	0
0	0	1	0	0	0

16.

NOT A AND (A OR (A OR B))

A	B	NOT A	A OR B	A OR (A OR B)	NOT A AND (A OR (A OR B))
1	1	0	1	1	0
0	1	1	1	1	1
1	0	0	1	1	0
0	0	1	0	0	0

17.

((B AND NOT NOT B) OR (A OR NOT NOT A)) OR (A AND NOT B)

A	B	NOT A	NOT B	NOT NOT A	A AND NOT B	NOT NOT B	(B AND NOT NOT B)	(A OR NOT NOT A	(B AND NOT NOT B) OR (A OR NOT NOT A)	((B AND NOT NOT B) OR (A OR NOT NOT A)) OR (A AND NOT B)
1	1	0	0	1	0	1	1	1	1	1
0	1	1	0	0	0	1	1	0	1	1
1	0	0	1	1	1	0	0	1	1	1
0	0	1	1	0	0	0	0	0	0	0

18.

(A OR B) OR (B AND A) AND (A AND B)

A	B	A OR B	B AND A	(A OR B) OR (B AND A)	(A OR B) OR (B AND A) AND (A AND B)
1	1	1	1	1	1
0	1	1	0	1	0
1	0	1	0	1	0
0	0	0	0	0	0

19.

(A AND B) OR (B AND A) AND (A OR B)

A	B	A OR B	B AND A	(A AND B) OR (B AND A)	(A AND B) OR (B AND A) AND (A OR B)
1	1	1	1	1	1
0	1	1	0	0	0
1	0	1	0	0	0
0	0	0	0	0	0

20.

(A AND B) OR (B AND NOT B) AND (A AND NOT NOT A)

A	B	NOT A	NOT B	NOT NOT A	A AND B	B AND NOT B	(A AND B) OR (B AND NOT B)	(A AND NOT NOT A	(A AND B) OR (B AND NOT B) AND (A AND NOT NOT A)
1	1	0	0	1	1	0	1	1	1
0	1	1	0	0	0	0	0	0	0
1	0	0	1	1	0	0	0	1	0
0	0	1	1	0	0	0	0	0	0

21.

(A AND B) AND (B AND NOT B) AND (A AND NOT NOT A)

A	B	NOT A	NOT B	NOT NOT A	A AND B	B AND NOT B	(A AND B) AND (B AND NOT B)	(A AND NOT NOT A	(A AND B) AND (B AND NOT B) AND (A AND NOT NOT A)
1	1	0	0	1	1	0	0	1	0
0	1	1	0	0	0	0	0	0	0
1	0	0	1	1	0	0	0	1	0
0	0	1	1	0	0	0	0	0	0

22.

(A AND B) OR (B AND NOT B) OR (B AND NOT NOT B)

A	B	NOT A	NOT B	NOT NOT A	A AND B	B AND NOT B	(A AND B) OR(B AND NOT B)	(B AND NOT NOT B)	(A AND B) OR (B AND NOT B) OR (B AND NOT NOT B)
1	1	0	0	1	1	0	1	1	1
0	1	1	0	0	0	0	0	1	1
1	0	0	1	1	0	0	0	0	0
0	0	1	1	0	0	0	0	0	0

23.

(A OR B) AND (B OR NOT B) OR (B OR NOT NOT B)

A	B	A OR B	NOT B	B OR NOT B	NOT NOT B	B OR NOT NOT B	(A OR B) AND (B OR NOT B)	(A OR B) AND (B OR NOT B) OR (B OR NOT NOT B)
1	1	1	0	1	1	1	1	1
0	1	1	0	1	1	1	1	1
1	0	1	1	1	0	0	1	1
0	0	0	1	1	0	0	0	1

24.

NOT ((A AND A) AND (B OR B)) OR NOT (A AND B)

A	B	(A AND A)	B OR B	((A AND A) AND (B OR B))	NOT ((A AND A) AND (B OR B))	A AND B	NOT (A AND B)	(NOT ((A AND A) AND (B OR B))) OR NOT (A AND B)
1	1	1	1	1	0	1	0	0
0	1	0	1	0	1	0	1	1
1	0	1	0	0	1	0	1	1
0	0	0	0	0	1	0	1	1

25.

NOT ((A OR A) AND (B OR B)) OR NOT (A OR B)

A	B	(A OR A)	B OR B	((A OR A) AND (B OR B))	NOT ((A OR A) AND (B OR B))	A OR B	NOT (A OR B)	NOT ((A OR A) AND (B OR B)) OR NOT (A OR B)
1	1	1	1	1	0	1	0	0
0	1	0	1	0	1	1	0	1
1	0	1	0	0	1	1	0	1
0	0	0	0	0	1	0	1	1

26.

(NOT A OR A) AND (B OR NOT B) OR NOT (A AND B)

A	B	NOT A	NOT B	NOT A OR A	B OR NOT B	(NOT A OR A) AND (B OR NOT B	(A AND B)	NOT (A AND B)	(NOT A OR A) AND (B OR NOT B) OR NOT (A AND B)
1	1	0	0	1	1	1	1	0	1
0	1	1	0	1	1	1	0	1	1
1	0	0	1	1	1	1	0	1	1
0	0	1	1	1	1	1	0	1	1

27. NOT (NOT A OR A) AND (B OR NOT B) OR NOT A

A	B	NOT A	NOT B	NOT A OR A	B OR NOT B	NOT (NOT A OR A)	NOT (NOT A OR A) AND (B OR NOT B	NOT A	NOT (NOT A OR A) AND (B OR NOT B) OR NOT A
1	1	0	0	1	1	0	0	0	0
0	1	1	0	1	1	0	0	1	1
1	0	0	1	1	1	0	0	0	0
0	0	1	1	1	1	0	0	1	1

28.

(B OR NOT A) AND (A OR NOT A) AND (B OR NOT B)

A	B	NOT A	NOT B	A OR NOT A	B OR NOT A	B OR NOT B	(B OR NOT A) AND (A OR NOT A)	(B OR NOT A) AND (A OR NOT A) AND (B OR NOT B)
1	1	0	0	1	1	1	1	1
0	1	1	0	1	1	1	1	1
1	0	0	1	1	0	1	0	0
0	0	1	1	1	1	1	1	1

29.

(A OR NOT NOT B) OR (A OR NOT NOT A) AND (A AND B)

A	B	NOT A	NOT B	NOT NOT A	A AND B	NOT NOT B	(A OR NOT NOT B)	(A OR NOT NOT A	(A OR NOT NOT B) OR (A OR NOT NOT A)	(A OR NOT NOT B) OR (A OR NOT NOT A) AND (A AND B)
1	1	0	0	1	1	1	1	1	1	1
0	1	1	0	0	0	1	1	0	1	0
1	0	0	1	1	0	0	1	1	1	0
0	0	1	1	0	0	0	0	0	0	0

30.

(A OR NOT NOT B) OR (A OR NOT NOT A) OR (NOT A AND B)

A	B	NOT A	NOT B	NOT NOT A	NOT A AND B	NOT NOT B	(A OR NOT NOT B)	(A OR NOT NOT A	(A OR NOT NOT B) OR (A OR NOT NOT A)	(A OR NOT NOT B) OR (A OR NOT NOT A) OR (A AND B)
1	1	0	0	1	0	1	1	1	1	1
0	1	1	0	0	1	1	1	0	1	1
1	0	0	1	1	0	0	1	1	1	1
0	0	1	1	0	0	0	0	0	0	0

31.

$(A + \sim B)(A + B) + (\sim A) + \sim A = AA + AB + (\sim B)A + (\sim B) B + \sim A$

$= A + A(B + \sim B) + 0 + \sim A$

$= A + A + \sim A = A + \sim A = 1$

USE FORMULAE

$\{\sim BB =0 , B + \sim B = 1 A + A = A , A + \sim A = 1\}$

32. $\sim A\sim B (((A * B) * (A * B))+ (A * B))$

$= \sim A\sim B (((A + B) * (A + B)) + A + B)$

$= \sim A\sim B ((AA + AB + BA + BB) + A + B)$

$= \sim A\sim B (A + A + B + B + A + B + A + B)$

$= \sim A\sim B (A + B)$

$= \sim A \sim B A + \sim A\sim BB = 0 + 0 = 0$

33. $A (\sim AB + B) + \sim A\sim B = A \sim A B + AB + \sim A\sim B$

$= 0B + AB + \sim A\sim B$

$= AB + \sim A\sim B$

$= A + B + \sim A + \sim B$

$= 1 + 1 = 1$

34. $\sim[\sim B (\sim AB + B) + \sim A (A + \sim A) + A] = \sim [\sim B\sim A B + \sim BB + \sim AA + \sim A\sim A + A]$

$= \sim [0B + 0 + 0 + \sim A + A]$

$= \sim [A + \sim A \} = \sim 1 = 0$

35. $\sim A[A(\sim A + B) + A .\sim B] = \sim A [A . \sim A + A . B + A . \sim B]$

$= \sim A [0 + A (B + \sim B)]$

$= \sim A [A (1)]$

$= \sim A A = 0$

36. B (A + ~B) + ~A B + ~ B

= BA + B~B + ~A B + ~B = BA + 0 + ~AB + ~B
= B (A + ~A) + ~B
= B + ~B = 1

37 ~ [A (B + ~B) (A + ~A)] + ~B = ~ [A(1)(1)] + ~B

= ~A + ~B = ~(AB)

38. BA + A(A + ~B) + ~AB + AB = BA + AA + A~B + ~AB + AB

= BA + A + A~B + B (~A + A)
= BA + A + A~B + B (1)
= A(B + ~B) + A + B
= A (1) + A + B = A + A + B = A + B = AB

39. B [~B~A + ~A(A + ~B) + A~B] = B [~B~A + ~AA + ~A~B + A~B]

= B [~B ~A + 0 + A~B]= B [~B (~A + A)] = B ~B (1)
= 0 = FALSE

40. Is the below statement TRUE or FALSE?
A [~A ~B + B (~A + ~B)] = A [~A ~B + B ~A + B ~B] = A [~A (~B + B) + 0]
= A ~A = 0 = FALSE

41. B (A + ~B) (~A + ~B) = B (A~A + A~B + ~B~A + ~B.~B)

= B (0 + A ~B + ~B ~A + 0)

= B (~B (A + ~A))

= B . ~B = 0

42. (~A + B) (A + ~B) = (~AA + ~A~B + BA + B.~B)

= (0 + ~A ~B + B A + 0)

= (~A + ~B + B + A)

= 1 + 1 = 1 = TRUE

43. ~ (~A + ~B) (A + B) = ~ (~AA + ~AB + ~BA + ~B.B)

= ~ (0 + ~A B + ~B A + 0)

= ~ (~A + B + ~B + A)

= ~ (1 + 1) = ~ 1 = 0 = FALSE

44. B[~B (B + ~B) (A + ~A)] + (~~A)(~A) = B[[~B. (1) (1)] + (~ ~A) (~A)
= B.~B + A.~A = 0 + 0 = 0

45.
~ (((A * B) * (A * B)) + (A * B)) + ~AB + A
= ~ (((A + B) * (A + B)) + A + B) + ~AB + A
= ~ ((AA + AB + BA + BB) + A + B) + ~AB + A
= ~ (A + A + B + B + A + B + A + B) + ~AB + A

= ~(A + B) + ~AB + A
= ~A * ~ B + ~AB + A = ~A(~B + B) + A = ~A + A =1

MATH KNOTS

1) Using the first letters of the names , Draw a graph relating to their meetups.
Amy can go to Bob's house , Amy can go to Cathy's house ,
Amy can go to Ella's house , Amy can go to Geeth's house ,
Geeth can go to Fiona's house , Fiona can go to Ella's house ,
Ella can go to Dan's house , Bob can go to Cathy's house ,
and Cathy can go to Dan's house.

Solution :

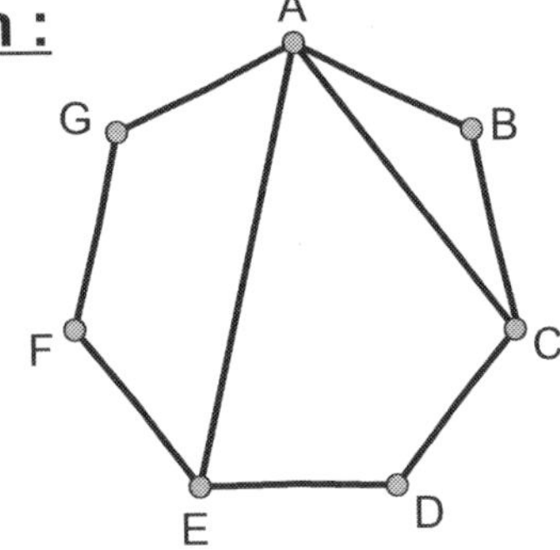

2) Using the first letters of the names , Draw a graph relating to their meetups.
Amy can go to Bob's house , Amy can go to Cathy's house ,
Amy can go to Ella's house , Amy can go to Geeth's house ,
Geeth can go to Fiona's house , Fiona can go to Ella's house ,
Ella can go to Dan's house , Bob can go to Cathy's house ,
Cathy can go to Dan's house , Geeth can go to Cathy's house ,
and Fiona can go to Cathy's house

Solution :

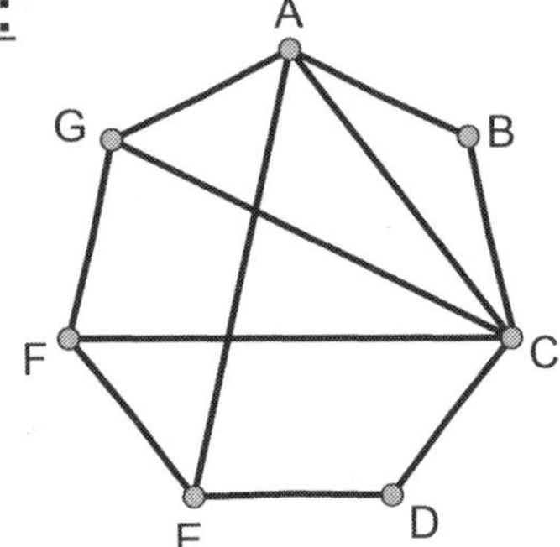

3) Using the first letters of the names , Draw a graph relating to their meetups.
Amy can go to Bob's house , Amy can go to Cathy's house ,
Amy can go to Ella's house , Amy can go to Geeth's house ,
Geeth can go to Fiona's house , Fiona can go to Ella's house ,
Ella can go to Dan's house , Bob can go to Cathy's house ,
Cathy can go to Dan's house , Geeth can go to Cathy's house ,
Fiona can go to Cathy's house , and Dan can go to Geeth's house

Solution :

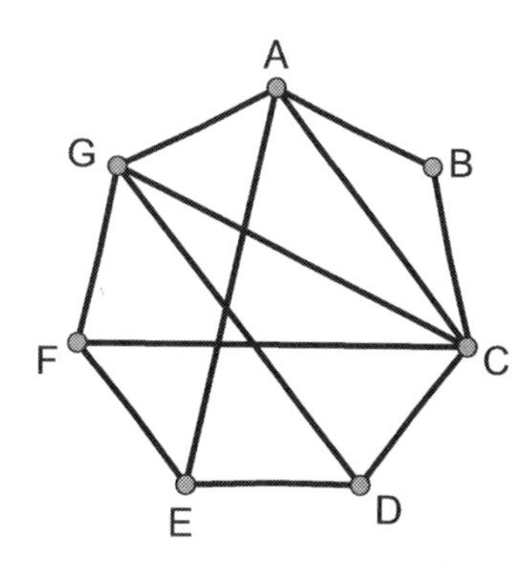

4) Zoom Zoom carnival has six crazy rides. There are only few paths to reach to one ride to another ride. Jade can go from ride A to E , ride A to B , ride A to F , ride A to D , ride F to B , ride F to E , ride D to E , ride D to C and ride B to C. Using the names of the rides , Draw a graph relating to their paths.

Solution :

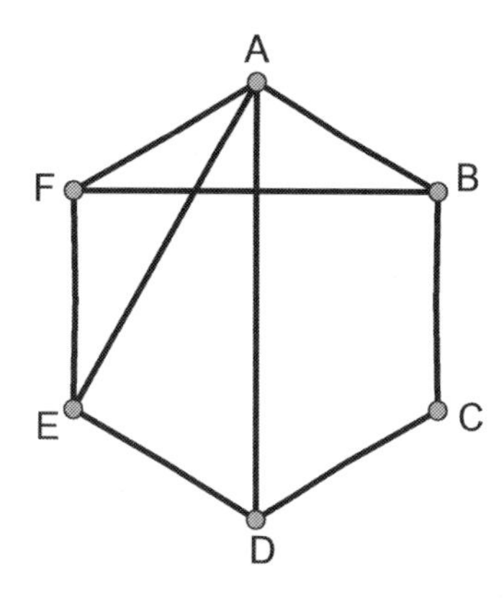

5) Zoom Zoom carnival has six crazy rides. There are only few paths to reach from one ride to another ride. Jade can go from ride A to B , ride A to F , ride E to C , ride F to B , ride F to E , ride D to E , ride D to C and ride B to C.
Using the names of the rides , Draw a graph relating to their paths.

Solution :

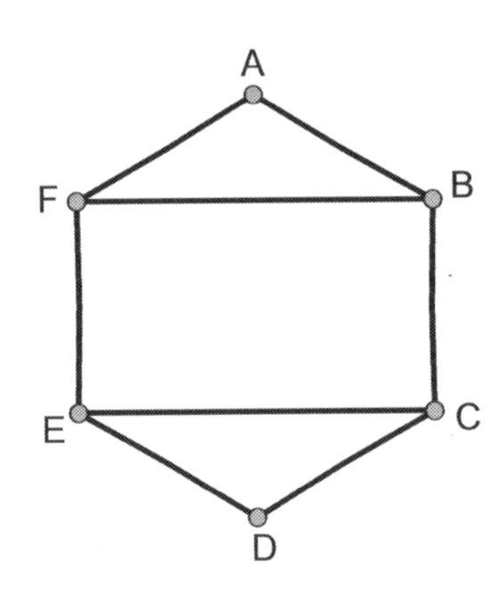

6) Sam's garden has six big apple trees named as A,B,C,D,E,F. Apple trees are connected to each other by a stone path. Sam can go from A to B , A to F , A to D , B to C , B to E , B to F , C to D , C to E , F to C , E to D and E to F.
Using the names of the trees , Draw a graph relating to the connecting stone paths.

Solution :

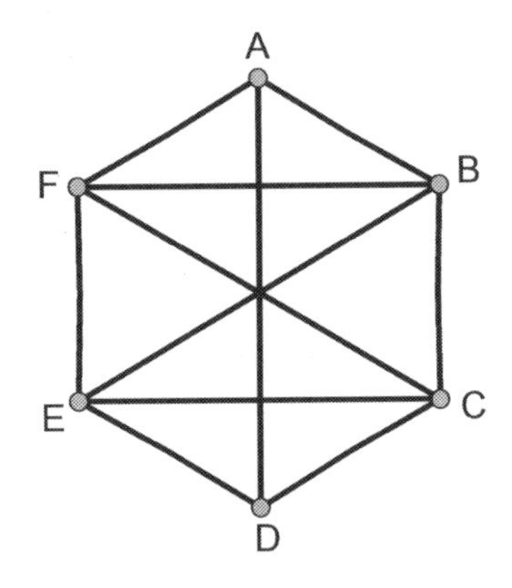

7) A bus leaves from Stop A to C , Stop A to D , Stop B to A , Stop D to B , Stop C to B and Stop D to C
Using the names of the stops , Draw a graph relating to the bus routes.

Solution :

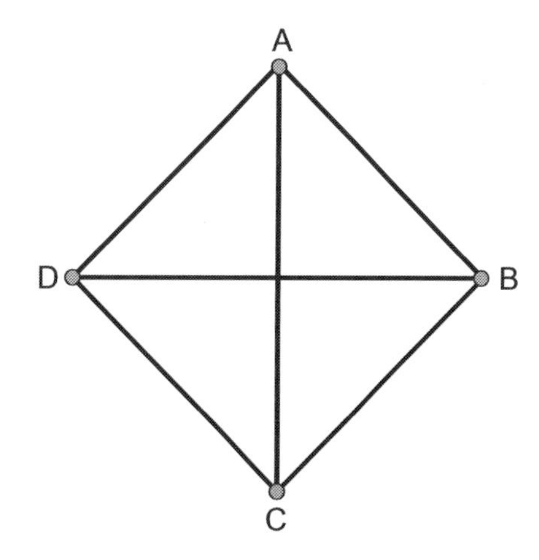

8) Island Oxygen has three bridges connecting to the city on the shore.
There are roads connecting to the bridges A,B,C. The roads and bridges are connected as A to O , O to C , B to O , A to C , B to C , and B to A.
Using the names of the roads and bridges , Draw a graph relating to their paths.

Solution :

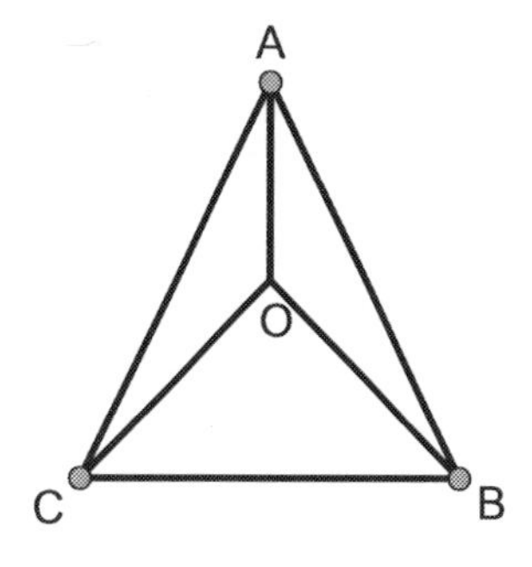

9) Island Oxygen has three bridges connecting to the city on the shore. There are roads connecting to the bridges A,B,C. The roads and bridges are connected as A to O , O to C , B to O , A to C , and B to A. Using the names of the roads and bridges , Draw a graph relating to their paths.

Solution :

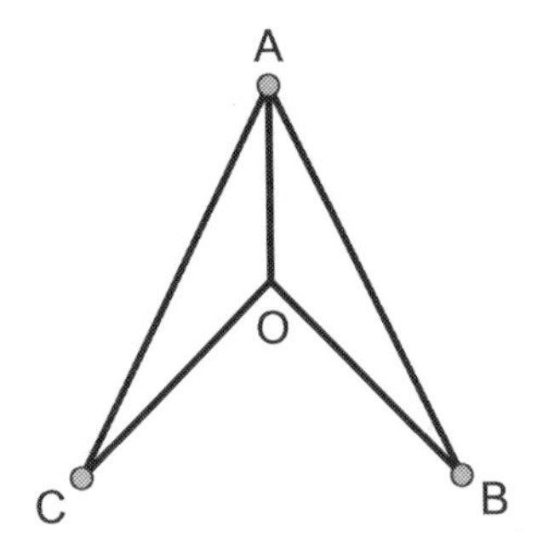

10) Draw an undirected graph with the vertices A , B , C , D , E , F and edges as AB , CA , BC , DE , DF , FE , EA , CD , and BF.

Solution :

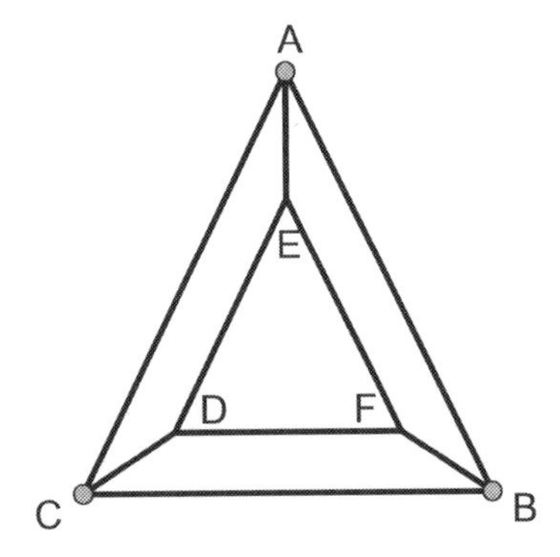

11) Draw an undirected graph with the vertices A , B , C , D , E , F and edges as AD , AE , DE , DC , BE , FD , EF , CF , and BF.

Solution :

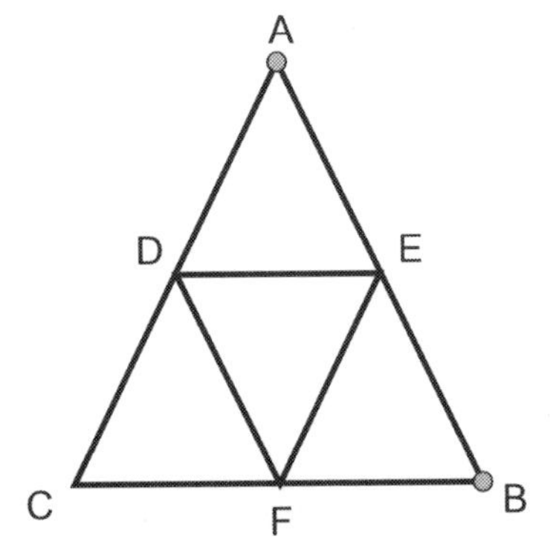

12) Draw an undirected graph with the vertices A , B , C , D , E , F and edges as AB , AD , EA , DB , FB , BC , CA , CF , FE , and ED.

Solution :

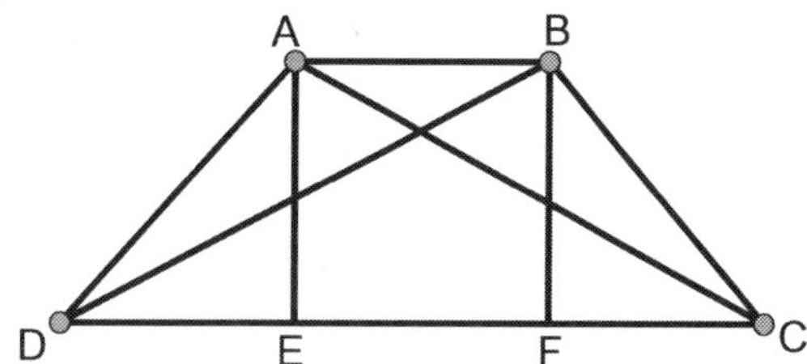

13) Draw an undirected graph with the vertices A , B , C , D , F and edges as AB , AD , DB , FB , BC , CF , and FD.

Solution :

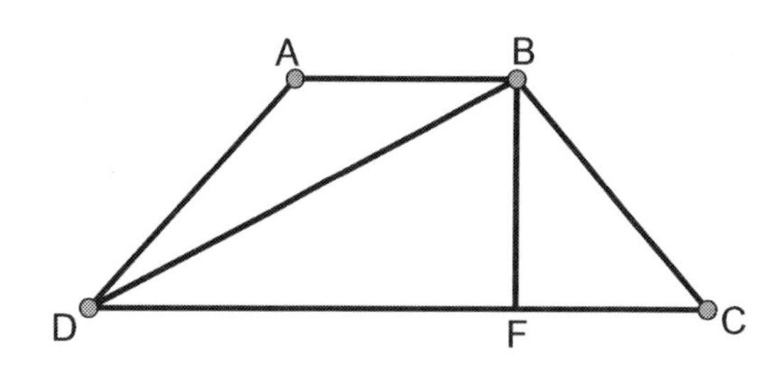

14) Draw an undirected graph with the vertices A , B , C , D , F and edges as AB , AD , DB , FB , BC , CA , CF , and FD.

Solution :

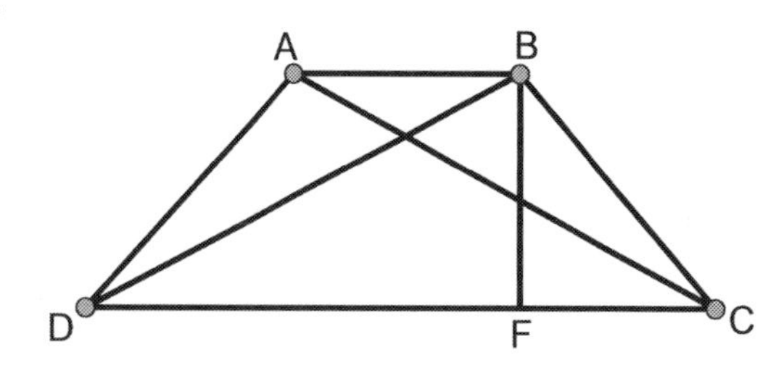

15) Draw an undirected graph with the vertices A , B , C , D , E , F and edges as AB , AD , EA , DB , FB , BC , CF , FE , and ED.

Solution :

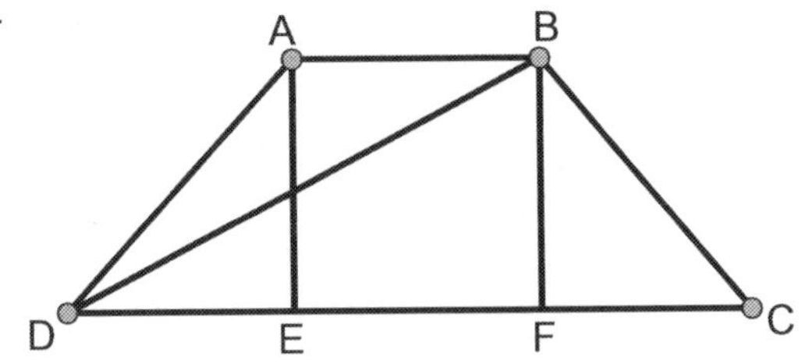

16) Draw an undirected graph with the vertices A , B , C , D , E , F and edges as AB , AD , EA , FB , BC , CF , FE , and ED.

Solution :

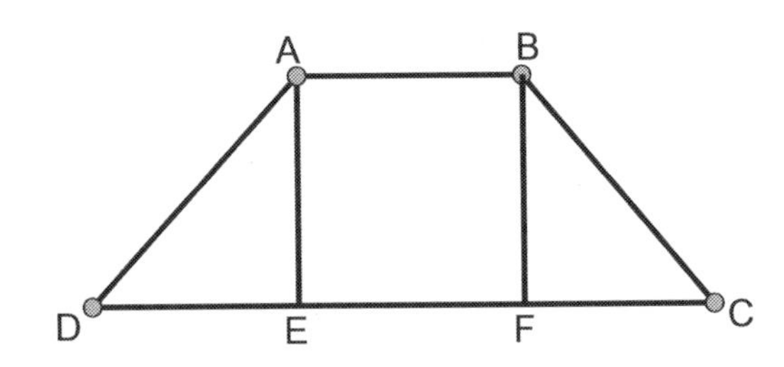

17) Draw an undirected graph with the vertices A , B , C , D , E and edges as AB , AD , EA , DB , BC , CE , and ED.

Solution :

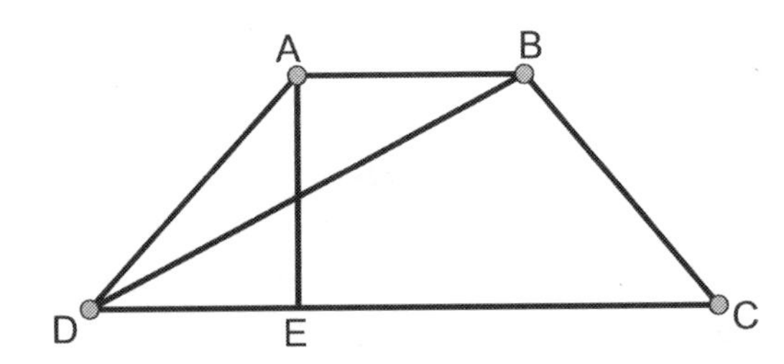

18) Draw an undirected graph with the vertices A , B , C , D , E and edges as AB , AD , EA , DB , BC , CA , CE , and ED.

Solution :

19) Draw an undirected graph with the vertices A , B , C , D , E , F , G and edges as AB , AG, FG , EF , ED , CD , CB.

Solution :

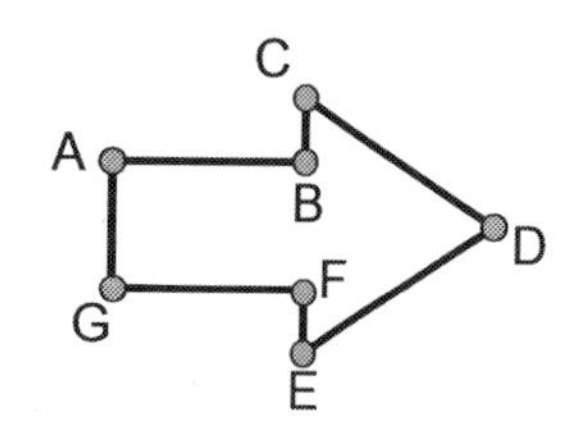

20) How many path of lengths equal to 2 exist in the below figure ? List all of them.

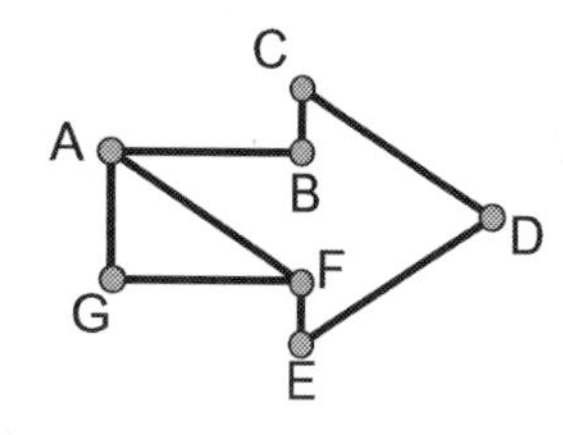

Solution :

ABC , BCD , CDE , DEF , EFG , FGA , GAB , AFE , AFG , FAB

21) How many path of lengths equal to 2 exist in the below figure ? List all of them.

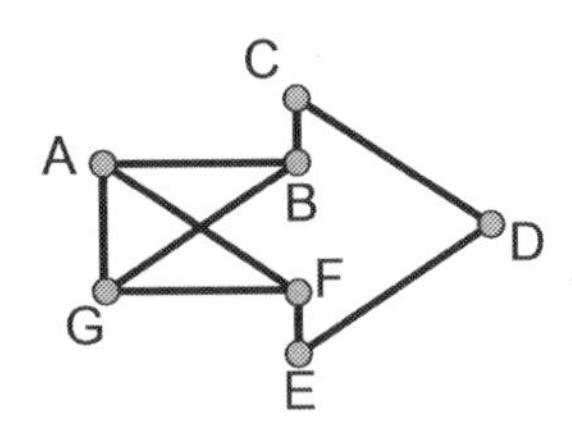

Solution :

ABC , BCD , CDE , DEF , EFG , FGA , GAB , AFE , AFG ,
GBC , GBA , BGF , FAB

22) How many path of lengths equal to 2 exist in the below figure ? List all of them.

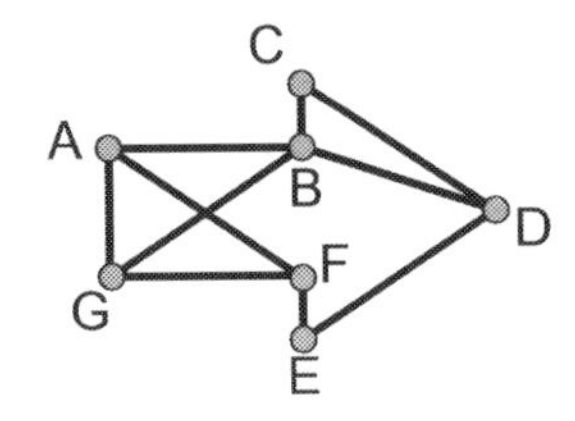

Solution :

ABC , BCD , CDE , DEF , EFG , FGA , GAB , AFE , AFG , GBC ,
GBA , BGF , BDE , FAB , ABD

23) List all cycles of length 3 of the below figure.

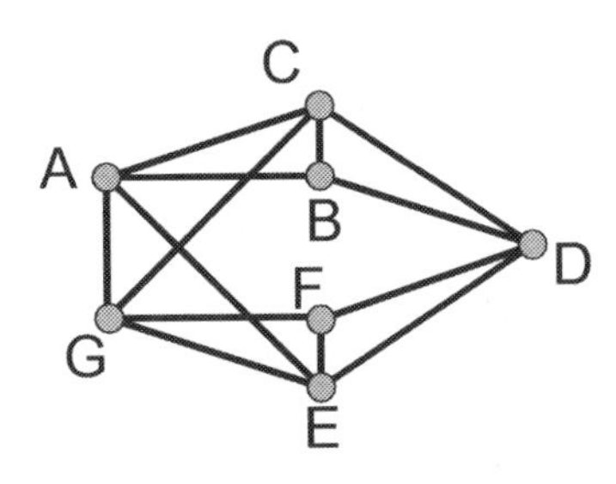

Solution :

ACGA , ACBA , AEGA , CDBC , DEFD , EFGE
The below paths are same as above
CBAC , BACB , BCDB , DBCD , EFDB , FEGF , FEDF , GAEG , GFEG , EAGE

24) List all cycles of length 3 of the below figure.

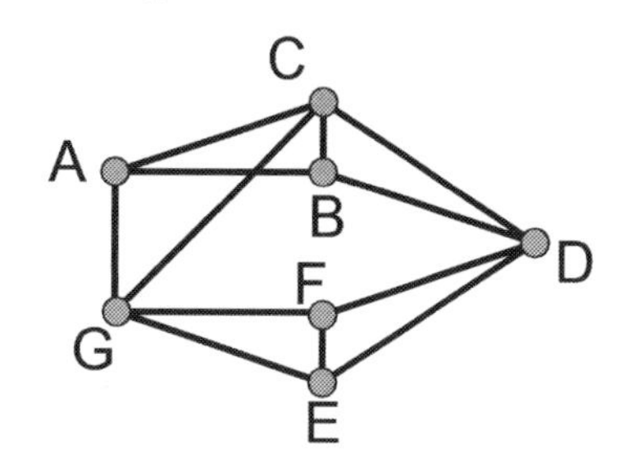

Solution :

ACGA , ACBA , AEGA , CDBC , DEFD , EFGE , CBAC
The below paths are same as above
BACB , BCDB , DBCD , EFDB , FEGF , FEDF , GFEG

25) List all cycles of length 3 of the below figure.

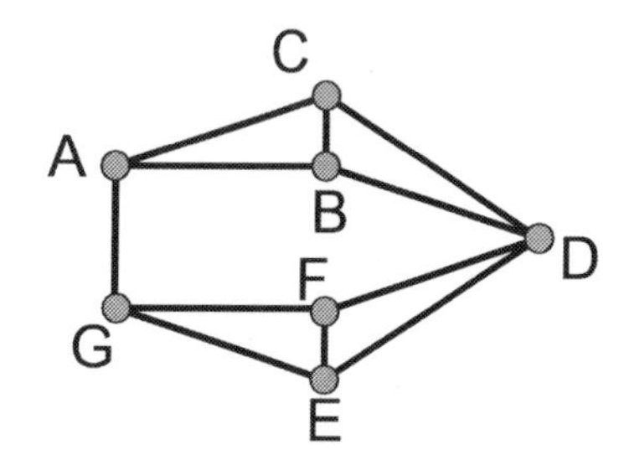

Solution :

ACBA , CDBC , DEFD , EFGE

26) Find if the following graph is traversable or not ?

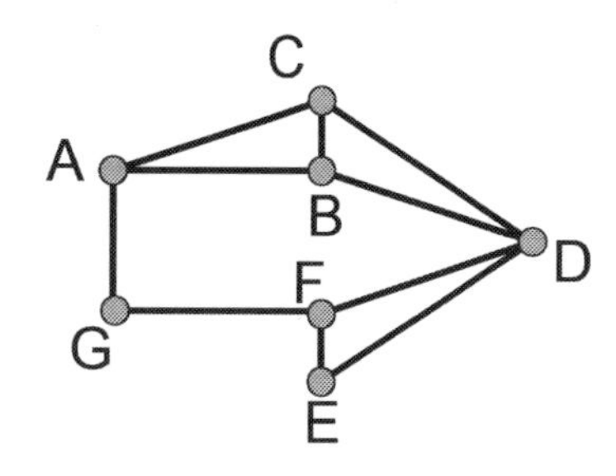

Solution :

The degree of each vertex is given below :
A = 3 , B = 3 , C = 3 , D = 4 , E = 2 , F = 3 , G = 2
There are 4 odd vertices. So ,the graph is not traversable.

27) Find if the following graph is traversable or not ?

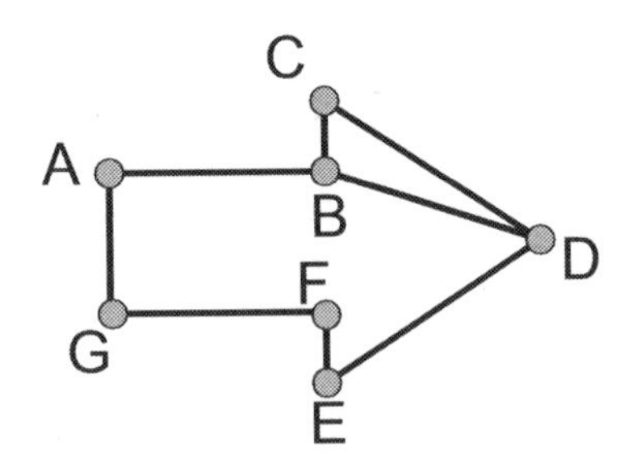

Solution :

The degree of each vertex is given below :
A = 2 , B = 3 , C = 2 , D = 3 , E = 2 , F = 2 , G = 2
There are 2 odd vertices. So ,the graph is not traversable.

28) Find if the following graph is traversable or not ?

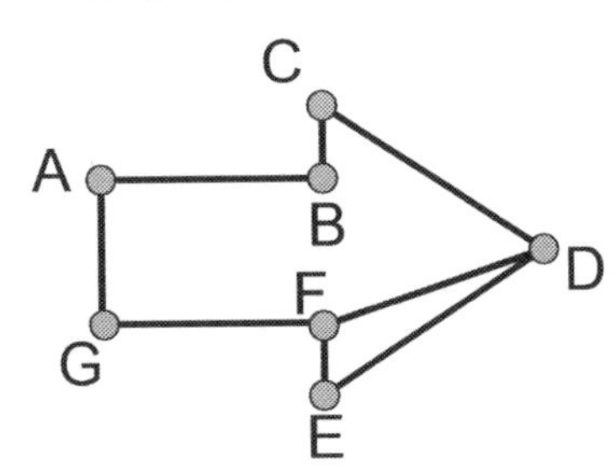

Solution :

The degree of each vertex is given below :
A = 2 , B = 2 , C = 2 , D = 3 , E = 2 , F = 3 , G = 2
There is 2 odd vertices. So ,the graph is not traversable.

29) Find if the following graph is traversable or not ?

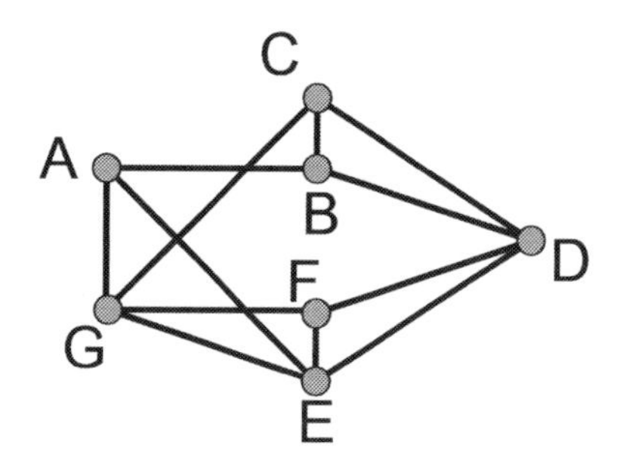

Solution :

The degree of each vertex is given below :
A = 3 , B = 3 , C = 3 , D = 4 , E = 4 , F = 3 , G = 4
There are 4 odd vertices. So ,the graph is not traversable.

30) Find if the following graph is traversable or not ?

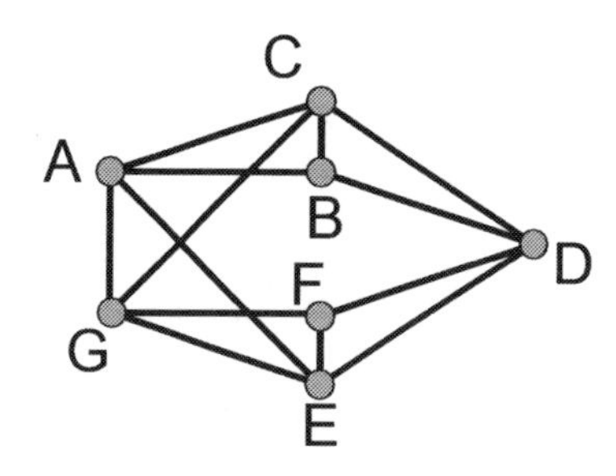

Solution :

The degree of each vertex is given below :
A = 4 , B = 3 , C = 4 , D = 4 , E = 4 , F = 3 , G = 4
There are 2 odd vertices. So ,the graph is traversable.

31) Find if the following graph is traversable or not ?

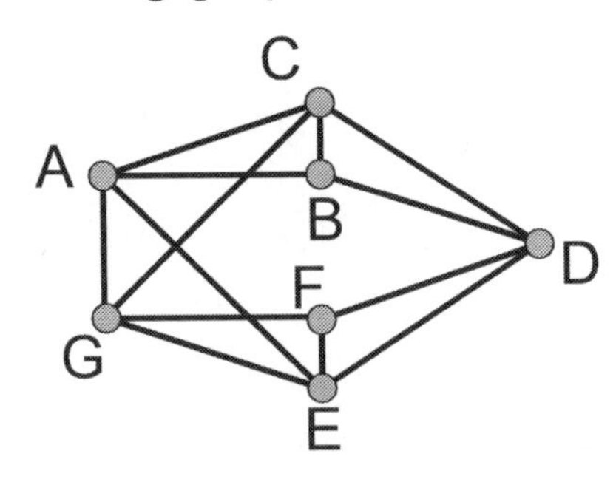

Solution :

The degree of each vertex is given below :
A = 4 , B = 3 , C = 4 , D = 4 , E = 3 , F = 3 , G = 4
There are 3 odd vertices. So ,the graph is not traversable.

32) List all cycles of length 2 of the below figure.

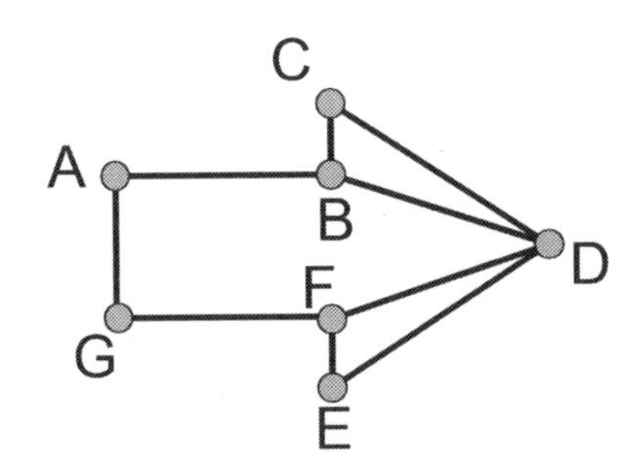

Solution :

ABC , ABD , AGF , BAG , BCD , BDF , BDE , CDF , CDE ,
DEF , DFG , EFG , CBD

33) List all cycles of length 2 of the below figure.

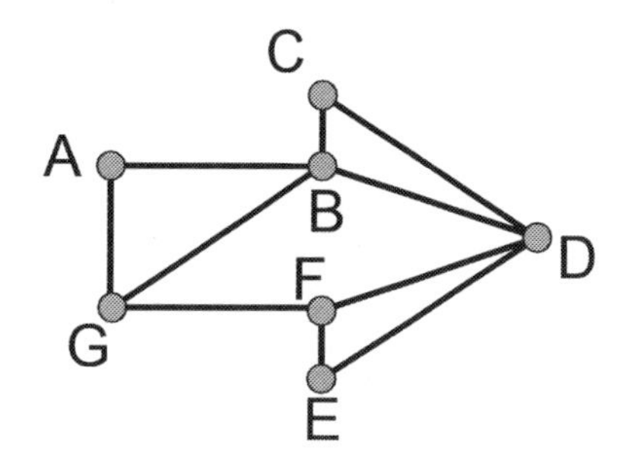

Solution :

ABC , ABD , AGF , BAG , BCD , BDF , BDE , CDF , CDE ,
DEF , DFG , EFG , BGF , GBD , GBC , CBD, DFE, AGB

34) List all cycles of length 2 of the below figure.

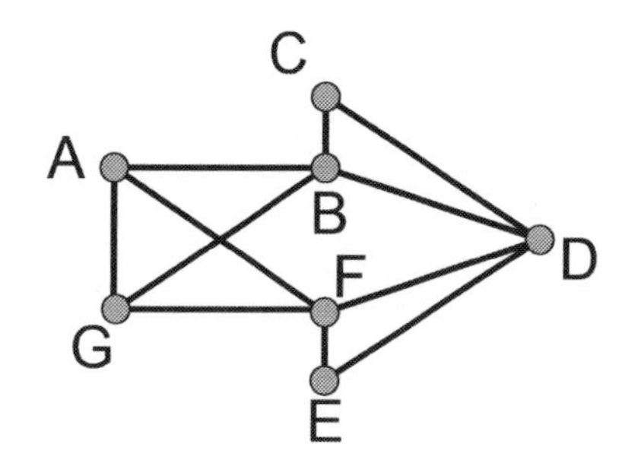

Solution :

ABC , ABD , AGF , BAG , BCD , BDF , BDE , CDF , CDE ,
DEF , DFG , EFG , BGF , GBD , AFD , AFE , DFA , GBC ,
AGB, CBD, EFD, GAF , FDE

35) List all cycles of length 2 of the below figure.

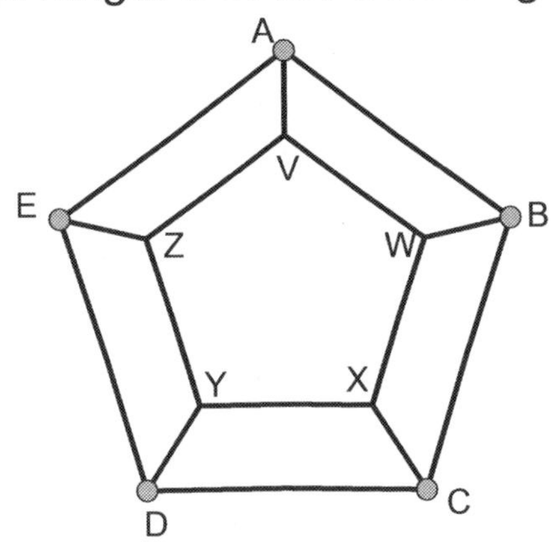

Solution :

ABC , BCD , CDE , DEA , XYZ , YZV , ZVW , VWX , AVZ ,
AVW , EZB , EZY , DYZ , DYX , CXY , CXW , BWX , BWV ,
WXY , AEZ , EAV , BAV , ABW , WBC , BCX , XCD , CDY , YDE , DEZ

36) List all cycles of length 3 of the below figure.

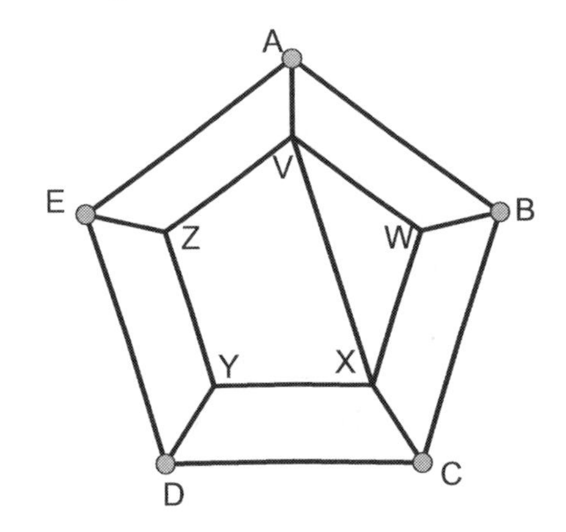

Solution :

ABC , BCD , CDE , DEA , XYZ , YZV , ZVW , VWX , AVZ ,
AVW , EZB , EZY , DYZ , DYX , CXY , CXW , BWX , BWV ,
AVX , CXV , XVW , VXY , XVZ , WXY , AEZ , EAV , BAV ,
ABW , WBC , BCX , XCD , CDY , YDE , DEZ , VXW

37) List all cycles of length 3 of the below figure.

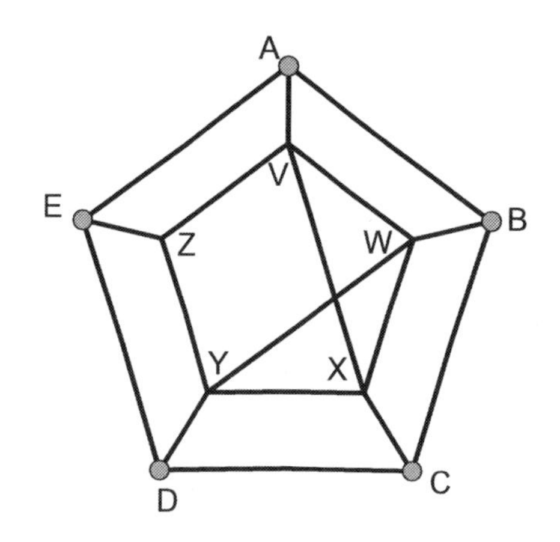

Solution :

ABC , BCD , CDE , DEA , XYZ , YZB , ZVW , VWX , AVZ , AVW , EZV , EZY , DYZ , DYX , CXY , CXW , BWX , BWV , AVX , CXV , XVW , VXY , XVZ , WXY , AEZ , EAV , BAV , ABW , WBC , BCX , XCD , CDY , YDE , DEZ , VXW , BWY , WYD , YWV , YWX , WYZ , WYX

38) Find if the following graph is traversable or not ?

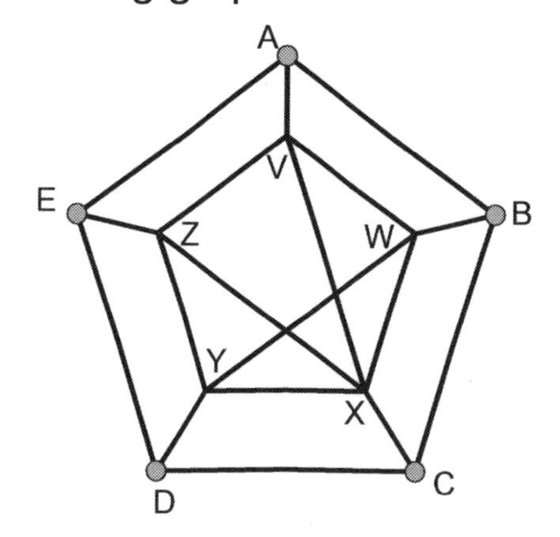

Solution :

The degree of each vertex is given below :
A = 3 , B = 3 , C = 3 , D = 3 , E = 3 ,
W = 4 , X = 5 , Y = 4 , Z = 4 , V = 4
There are 6 odd vertices. So ,the graph is not traversable.

39) Find if the following graph is traversable or not ?

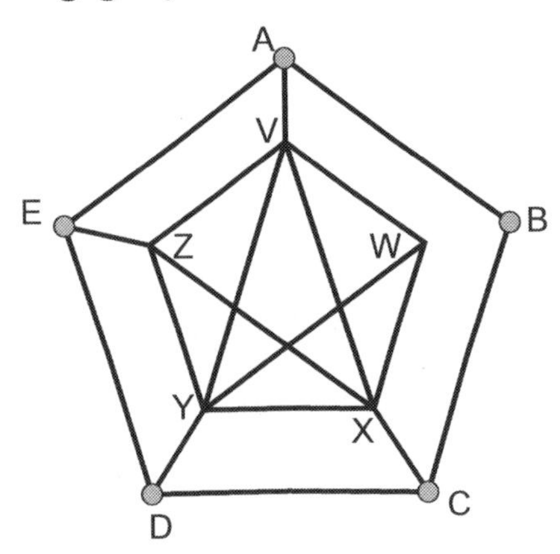

Solution :

The degree of each vertex is given below :
A = 3 , B = 2 , C = 3 , D = 3 , E = 3 ,
W = 3 , X = 5 , Y = 5 , Z = 4 , V = 5
There are 8 odd vertices. So ,the graph is not traversable.

40) Find if the following graph is traversable or not ?

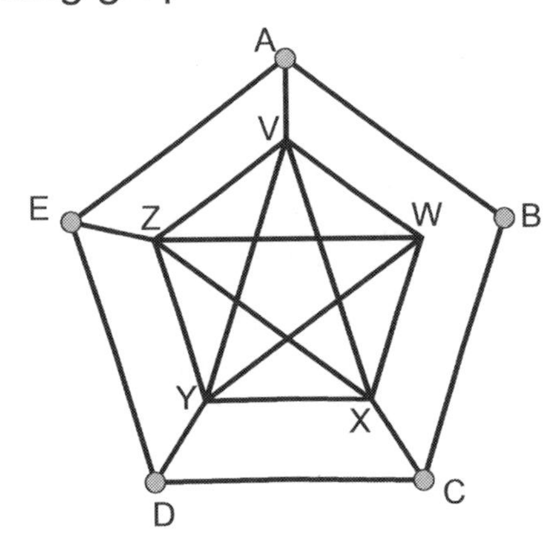

Solution :

The degree of each vertex is given below :
A = 3 , B = 2 , C = 3 , D = 3 , E = 3 ,
W = 4 , X = 5 , Y = 5 , Z = 5 , V = 5
There are 8 odd vertices. So ,the graph is not traversable.

41) Find if the following graph is traversable or not ?

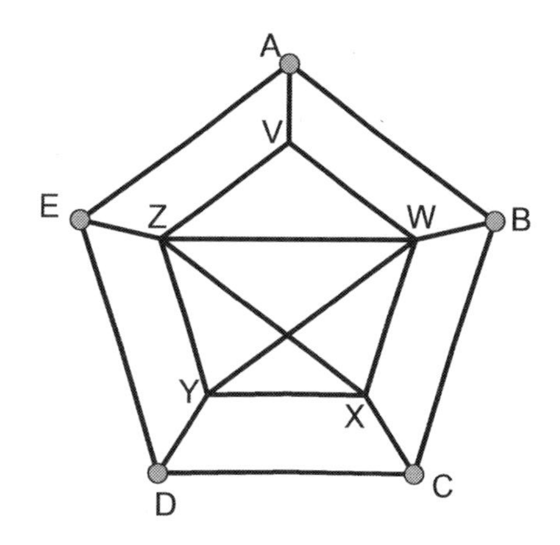

Solution :

The degree of each vertex is given below :
A = 3 , B = 3 , C = 3 , D = 3 , E = 3 ,
W = 5 , X = 4 , Y = 4 , Z = 5 , V = 3
There are 8 odd vertices. So ,the graph is not traversable.

42) Find if the following graph is traversable or not ?

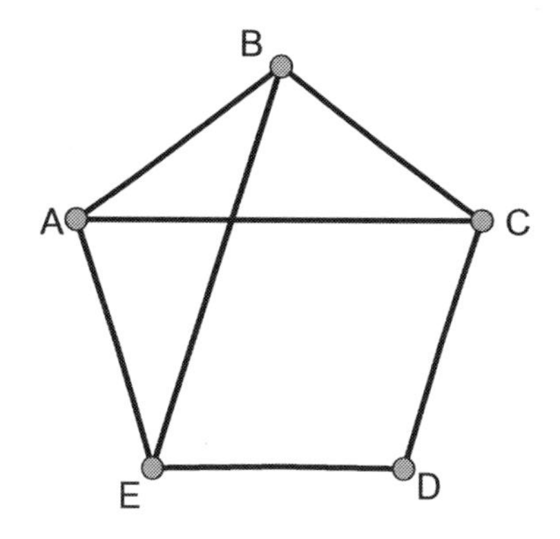

Solution :

The degree of each vertex is given below :
A = 3 , B = 3 , C = 3 , D = 2 , E = 3
There are 4 odd vertices. So ,the graph is not traversable.

43) Find if the following graph is traversable or not ?

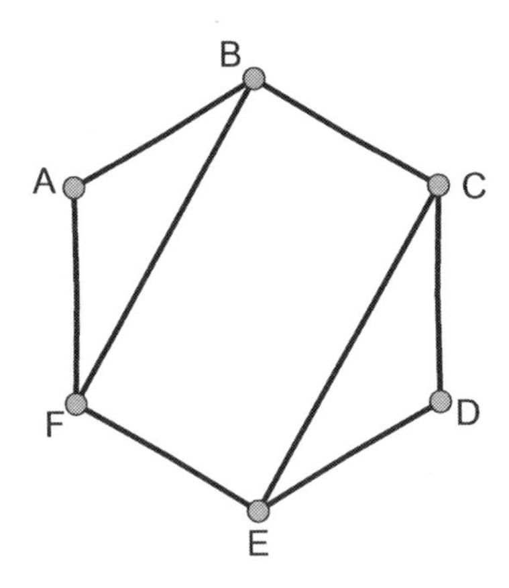

Solution :

The degree of each vertex is given below :
A = 2 , B = 3 , C = 3 , D = 2 , E = 3 , F = 3
There are 4 odd vertices. So ,the graph is not traversable.

44) Find if the following graph is traversable or not ?

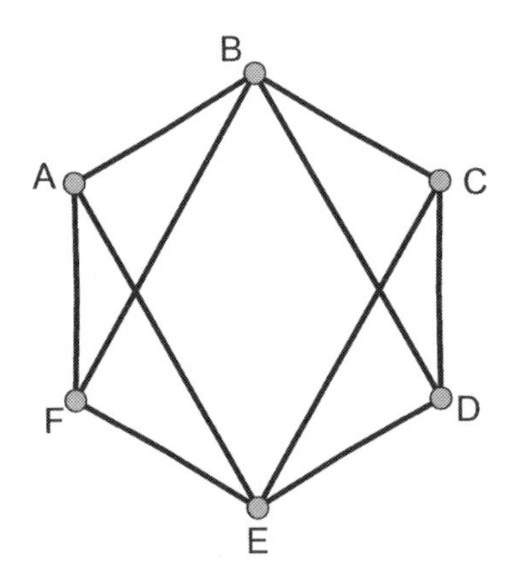

Solution :

The degree of each vertex is given below :
A = 3 , B = 4 , C = 3 , D = 3 , E = 4 , F = 3
There are 4 odd vertices. So ,the graph is not traversable.

45) Find if the following graph is traversable or not ?

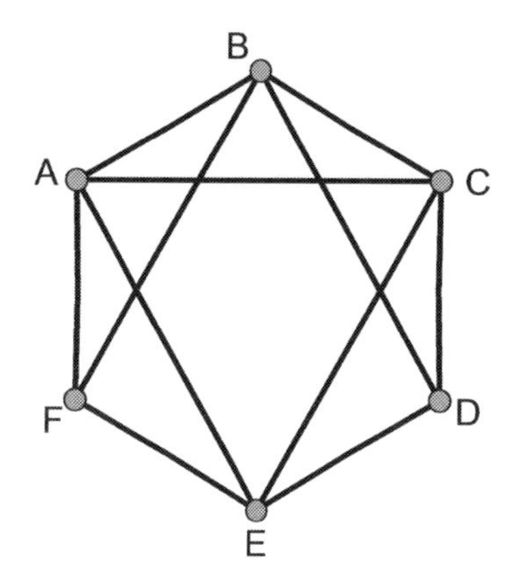

Solution :

The degree of each vertex is given below :
A = 4 , B = 4 , C = 4 , D = 3 , E = 4 , F = 3
There are 2 odd vertices. So ,the graph is not traversable.

Made in the USA
Columbia, SC
16 November 2024